T0237534

A METHOD OF
TEACHING CHEMISTRY
IN SCHOOLS

Slate Quarry at Llanberis.

A METHOD OF
TEACHING CHEMISTRY
IN SCHOOLS

BY

A. M. HUGHES, B.Sc. (London)
SCIENCE MISTRESS OF THE L.C.C. SECONDARY SCHOOL, ELTHAM

AND

R. STERN, B.Sc. (London)
SCIENCE MISTRESS OF THE NORTH LONDON COLLEGIATE SCHOOL

CAMBRIDGE:
at the University Press
1906

CAMBRIDGE
UNIVERSITY PRESS

University Printing House, Cambridge CB2 8BS, United Kingdom

Cambridge University Press is part of the University of Cambridge.

It furthers the University's mission by disseminating knowledge in the pursuit of
education, learning and research at the highest international levels of excellence.

www.cambridge.org
Information on this title: www.cambridge.org/9781107456525

© Cambridge University Press 1906

First published 1906
First paperback edition 2014

A catalogue record for this publication is available from the British Library

ISBN 978-1-107-45652-5 Paperback

PREFACE.

IN this little book we have endeavoured to work out a new method of teaching elementary chemistry in schools—a method based entirely upon the principle of working from the known to the unknown. Although not actually indicated in the book, it is intended that every experiment should be suggested and carried out by the pupils, the part of the teacher being only to guide and supervise. At the same time the teacher must reserve the right of selecting the experiment to be done by the class when several have been suggested, and, in this way, preventing time being wasted in trying experiments which would be of little value to the children and which would break the sequence of their work.

It will be found that by following this method of procedure, the class can work along the main lines suggested by us, although naturally the work must vary considerably in the hands of different teachers. When working along heuristic lines, every teacher knows that questions con-

stantly arise which the children cannot answer until they
have a wider knowledge of the subject. These questions
are generally put on one side at the time, with the result
that they are forgotten. We insist that the children
should write these questions in note-books kept in the
laboratory for the purpose, so that they can always be
referred to, and by being kept in this way constantly in
the memory of the children, these questions will prove
not only a valuable stimulus to further endeavour, but
also form a basis for future investigation when the
children know more about the subject. These question-
books will be found a source of great interest, and
will prevent much waste of time, as the children will
frequently, while waiting for a solution to evaporate or a
crucible to cool, try to answer some question in their
books to which they think they have found a clue.

Another point in this course we wish to emphasise is
that, starting from a few familiar substances in every-day
use, the children build up step by step their knowledge
of many chemical substances which they have themselves
prepared and of which they have found the properties.
Only in a very few instances are they given a new
substance, and even then it is introduced in connection
with their work from a historical standpoint, or because
it is used in a manufacturing process. In this way
the children are taught to realise that the science is
intimately connected with their every-day life. It must

be clearly understood that at first the experiments are not regarded entirely from the chemical standpoint, the object being mainly to cultivate powers of observation and dexterity in manipulation. At first also the children are allowed to recognise a substance from one or two properties; as their knowledge increases, however, they will themselves realise that several confirmatory tests are necessary. Although, from the beginning of the course, chemical nomenclature is used, the terms are names only to the children and convey no idea of composition. Symbols must not on any account be introduced, but at the end of the course the class will be ready to understand the principles underlying the atomic theory.

Throughout this course we have suggested that the children should keep a specimen of each substance they have themselves prepared, and that they should use these collections for purposes of identification and comparison.

It is advisable to keep these specimens in small test-tubes or specimen-tubes about two inches in length. To make them easy of access, and to prevent confusion, these tubes, carefully labelled, should be kept in racks fitted in convenient spaces on the laboratory walls, a rack being provided for each set of workers.

These racks can be made quite inexpensively of a piece of wood about two feet long and a foot wide, from which project ledges an inch wide at intervals of four inches. At $1\frac{1}{2}$ inches above each ledge there should be

a rail of stout wire, or a row of clips, to keep the tubes
in position and allow the labels to be seen.

A rack of this size will hold about one hundred
tubes, and the contents of the tubes can be viewed at
a glance. It will be found that the endeavour to get
good specimens for the collections will produce a spirit
of healthy rivalry in the class, and prove an excellent
stimulus to careful and painstaking work.

A. M. H.

R. S.

October, 1906.

CONTENTS.

CHAPTER I.

CHAPTER II.

CHAPTER III.

CHAPTER IV.

CHAPTER V.

PLATES.

CHAPTER I.

Examination of substances brought from home.
As an introduction to this course of Elementary Chemistry,
the pupils should be asked to bring specimens of solid
substances which are in common use at home. A list of
the materials brought should be written by the teacher
on the black-board. This list will probably include salt,
sugar, flour, rice, coal, wood, tea, coffee, washing-soda, sal-
ammoniac, saltpetre, alum, whitening, bath-brick, sulphur,
beetroot.

A description of each substance brought by the
children should be written in the class-book. The
effect of heating the substance in an open crucible
should then be tried, and the change of appearance, if
any, noted. From these experiments a classification of
the substances into those which char, and those which
do not char, will naturally suggest itself. By collecting
the class results, writing them on the blackboard, and
getting the children to suggest the origin of each substance, it will be discovered that those which char have
been derived from living matter, and hence are called
organic (organism, a living being); the other substances
on the list which are obtained from non-living matter are
termed *inorganic*.

This classified list should be entered into the note-book of every member of the class and kept for reference.

Fig. 1.

The children will also find that salt, bath-brick and whitening do not change in appearance when heated. Washing-soda and alum first liquefy, but on further heating leave a white solid residue in the crucible.

Nitre fuses and seems to boil, but solidifies when cooled. Sulphur melts, darkens in colour and burns with a blue flame, leaving no residue, while suffocating white fumes are given off. Sal-ammoniac sublimes.

Sugar melts, blackens and takes fire, and seems to disappear. Wood chars, burns, diminishes in quantity until finally only a small quantity of white ash is left, which does not change on continued heating. When beetroot is heated for some time, only a small quantity of white ash is left. This ash must be kept for future investigation.

When the coal is examined, some pieces will be found to contain gold-coloured flecks, and if these are removed with a penknife and heated separately, they will be found not to change unless heated strongly, although the coal itself takes fire and burns. These flecks are the substance called iron pyrites.

NOTE. It is advisable to select beetroot as the typical vegetable because the ash contains a large quantity of potassium salts.

Examination of building materials. The class must next be asked to bring specimens of materials used in building houses, and it will be found that an average class will bring the following substances:—sand, sandstone, flagstone, marble, slate, iron, zinc, lead, copper, brick and lime.

This list will naturally vary very much according to the locality of the school, and this part of the work is open to considerable variation. In many parts of England the children can now be made acquainted with ores, can be taken to see quarries and mines, and the work correlated with the geography lessons. In a London school, where there are many difficulties in getting specimens and going on expeditions, the substances can be studied in the following order:—

Sand. The appearance, colour and properties of the sand should be carefully noted, and then the localities in which it is found discussed. The children will all know that sand is found on the sea-shore, and by consideration of the other materials found there, such as pebbles and pieces of rock broken off the cliffs, they can be led

Sand-pit.

Sandstone Quarry.   Higher Bebington.
[*By kind permission of Mr Wells.*]

to suggest the origin of sand, *i.e.* that the rocks have
been worn away by the action of waves and weathering.

Further questioning will elicit the information that sand is also found inland, and that it is dug out of sand-pits which the children may have seen and of which photographs may be shown. The class will then be led to consider what would be the effect of pressure on such beds of sand, especially if the grains become cemented together, and to suggest that sandstone would be formed. Sandstone should then be examined to see if it does consist of sand particles cemented together, and pieces of paving-stone examined and compared with the sandstone. The localities from which sandstone and paving-stone are obtained should then be found out by the class, the children being allowed to consult books and ask questions at home.

At the next lesson the information brought by the children should be tabulated on the blackboard by the teacher and photographs or pictures of large sandstone and flagstone quarries should be shown, while each child should colour on a sketch-map the sandstone districts of England.

The sandstones and flagstones constitute a numerous family and as old shore sediments occur in almost every geological formation:—the Old Red Sandstone, Carboniferous; New Red Sandstone, Jurassic and Wealden.

For the most part they consist of grains of sand consolidated by pressure and cemented by silica, carbonate of lime, or oxide of iron. The localities from which sandstone is obtained for building purposes are very numerous; the following being a short list :—Carboniferous sandstones in many parts of Yorkshire, Derbyshire, Lancashire, Durham, Northumberland, Gloucestershire, Glamorganshire, etc. New Red Sandstones in Cheshire, Staffordshire and Worcestershire. The oolitic sandstone of Whitby and the Wealden sandstone of Tunbridge Wells are also used.

The following are some of the better-known sandstone quarries:—
Appleton (near Huddersfield), Darley Dale (Derbyshire), Forest of
Dean (Gloucestershire), Gazeby (Yorkshire), Mansfield (Notting-
hamshire), Matlock (Derbyshire). Much of the sandstone known
in London as "York stone," which is used for steps, landings, etc.,
comes from the South Owram Quarry in Yorkshire.

Lime. This substance is used with sand for making
mortar, and should therefore be examined next, but care
must be taken in handling it. Its appearance should be
noted by the class, as also the effects of adding water to
it, such as difference in temperature, absorption of water
and increase in bulk, thus getting the pupils to realise
the physical difference between "quicklime" and "slaked
lime."

The children should be encouraged to notice how
workmen prepare slaked lime and the mortar for building
purposes, and to write an account of what they have
seen. Most children will know that lime is made by
calcining limestone or chalk in a lime-kiln, so they can
examine pieces of limestone and chalk (whitening) and
compare them with lime, and if time permits they can
themselves try the effect of strongly heating a piece of
limestone to see whether they do get a substance formed
that behaves like lime when treated with water.

Marble can also at this stage be examined and described, but the class will not realise that there is any
connection between limestone, chalk and marble. The
children should be encouraged to find out from what
parts of England limestone, chalk and marble are
obtained, and should mark these districts in their
sketch-maps. Photographs of limestone quarries, marble
quarries, and chalk-pits or chalk cliffs should be shown.

Limestone Quarry at Llandulas.

The limestones are all of sedimentary origin, formed either from the remains of organisms or by mechanical deposition in deep sea or lake basins, or in certain cases by chemical precipitation. They vary in compactness from the softness of powdered chalk to the hardness of dense marble. The oolitic limestones are composed of spherical grains cemented together, and they are extensively used for building purposes. Almost all important buildings in London from the beginning of the 17th century until the present time have been built of Portland stone, one of the oolitic limestones. Limestones for building purposes are obtained from Carnarvonshire, Derbyshire, Devonshire, Dorsetshire, Gloucestershire, Northamptonshire, Rutlandshire, Somersetshire, Wiltshire, Yorkshire.

Lime, in addition to its use for making mortar, is used in the preparation of whitewash, which is formed of quicklime, mixed while hot with plenty of water.

Whiting, or whitening, is a name for levigated chalk which has not been calcined. In order to prepare this substance for use, six pounds of the powder is covered with water, left for six hours, and then mixed with a pound of double size and left to stand until it becomes like jelly. Later it is diluted with water and is then ready for use. Putty is made by mixing dried and finely-ground whiting with raw linseed oil.

Slate. Pieces of slate should next be examined by the children, the colour, lamellar fracture and relative softness noted.

When a piece of slate is breathed upon, a peculiar clayey odour is noticed; this property and the appearance of the slate will lead the children to infer that slate is hardened clay, and therefore hardened mud. Slate, however, has also been subjected to lateral pressure, so that it does not split along the planes of the original bedding.

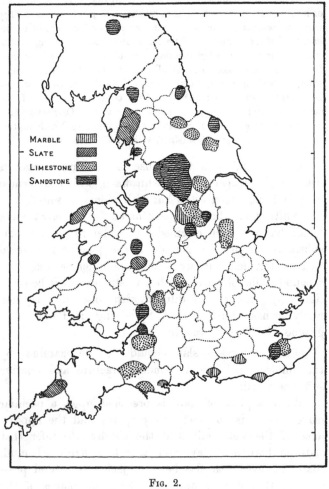

MARBLE
SLATE
LIMESTONE
SANDSTONE

FIG. 2.

Pant Brick Works, Ruabon.

9

The finest slates come from Wales, and the best known quarries are at Ffestiniog, Dinorwic, and the Penrhyn quarry.

English slates come from Westmoreland, Lancashire, Cornwall, Devonshire and Leicester.

Brick. Brick should next be examined, its preparation from clay discussed, and a photograph of a brickfield shown. The properties of clay, such as its appearance, plasticity, &c. should be noted, while the class might be led to suggest its formation from mud.

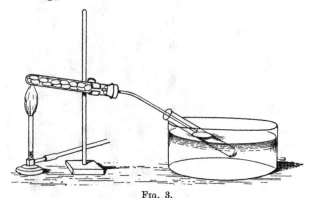

Fig. 3.

Bricks are made by baking clay. The class might try to make bricks from clay, baking their little bricks in the drying-oven and furnace. After it has been in the drying-oven the appearance of the clay is changed, and the class will suggest that this change has been caused by water having been given off. Then some clay might be heated in an evaporating dish, and a funnel or any cold surface held over the dish. Drops of a liquid will condense on it. In order to obtain a larger amount of this liquid, some small lumps of clay should be placed in a large test-tube fitted with a cork and delivery tube, leading into a test-tube surrounded by water, in which the liquid can be condensed as in Fig. 3. The test-tube containing the clay must be cautiously heated, and the mouth should be somewhat lower than the closed end or the water will trickle back and crack the tube.

The effect of heat on this liquid can be tried by

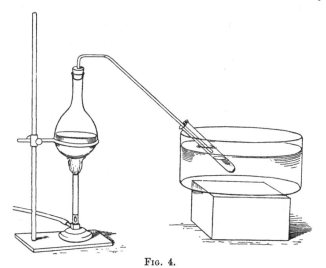

Fig. 4.

evaporating it in a watch-glass on a sand-bath, and it will be found that no residue is left. The children will want to find out whether water behaves in this way, and they will use tap-water for the purpose. On finding that a residue is left on the watch-glass, after a little thought they will suggest trying whether the water obtained by condensing the steam from tap-water will leave a residue.

A convenient way of obtaining this condensed steam is to place some water in a flask fitted with a cork and delivery tube, the latter passing into a test-tube surrounded with cold water as in the previous experiment.

This condensed steam evaporated on a watch-glass leaves no residue, so that the liquid obtained by heating clay behaves like condensed steam, and further investigation will show that its boiling point is the same as that of water, as also its freezing point.

Distillation. The process of converting water into

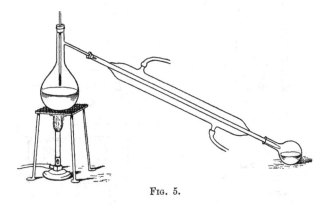

Fig. 5.

vapour and condensing this vapour is called *distillation* (from Latin *de* down and *stillo* I drop), and water so produced is called distilled water. To obtain distilled water more easily a Liebig's condenser can be used. Since water can be separated from solid impurities in this way, the question will be asked by the children, "Can it be separated from liquid impurities in the same way?" They may be allowed to try to separate a mixture of methylated spirit and water.

A simple way of doing this experiment is to put the mixture, which should contain a small percentage of methylated spirit, into a Wurtz distillation flask fitted with a cork having a hole through which a thermometer passes; the bulb of the thermometer must be above the liquid in the flask, and the tube of the flask should be connected with a Liebig's condenser.

It will be found that at 78° C. a liquid will distil over, which by its smell and inflammability can be recognised as methylated spirit. As the temperature rises the receiver should be replaced and the second distillate collected. This will smell like methylated spirit, but it will not burn. At 100° C. the receiver should be again changed, and now the distillate will be found to be water. Since pure water can be obtained from a mixture of water and a liquid, or from water and a solid, the question will arise—to what practical purposes can the distillation of water be applied? Amongst other things the preparation of drinking water from sea-water on board ship will be suggested, and this can be tried experimentally to see if it be possible. Some sea-water can be heated in the still, the distillate will prove to be drinkable. The residue in the flask can be put in an evaporating dish and evaporated to dryness. A saline solid is left, which can be redissolved in water and a clear solution obtained.

Solution. The class will want to find out if other solids will disappear when added to water, and salt, chalk, lime and iron pyrites can be tried. To find out if anything is dissolved the mixture can be filtered and the filtrate evaporated on a watch-glass.

If distilled water has been used the chalk and the iron pyrites filtrate give no residue, the lime filtrate a slight one. The chalk and iron pyrites are therefore insoluble, the lime has evidently a solubility limit. The salt is readily soluble, but, has it also a solubility limit? This can be tried by taking 50 c.c. of distilled water in a

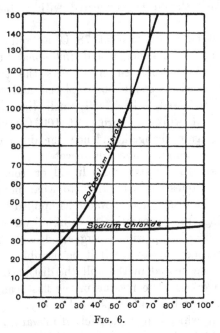

Fig. 6.

beaker and adding salt gradually, stirring the mixture meanwhile with a glass rod until some salt is left at the bottom undissolved. By means of a pipette 25 c.c. of this clear solution are placed in a weighed dish and evaporated to dryness; when cool the dish should be re-weighed and the amount of salt, which can be dissolved in 100 c.c. of water at the temperature of the air, found. The same experiment can be tried by other members of the class with water at different temperatures and the results collected and tabulated on the blackboard and the solubility curve of salt drawn.

NOTE. To do this experiment it is necessary to warm the pipette; this is best done by attaching a piece of rubber tubing to the end and sucking up hot water from a beaker.

The solutions must be saturated, and the temperature kept constant for some time before extracting a sample of the solution.

In the same way, the solubility curve of nitre should be found. It has been seen that the solubility of these substances varies according to the temperature of the water used: chalk and iron pyrites are insoluble even in hot water.

The children at this stage will want to find the effect of first heating the chalk or iron pyrites and of then trying the action of water on them. Some powdered chalk should be placed in a crucible and strongly heated in a muffle furnace or in a table furnace for 15 or 20 minutes. When cool, the substance in the crucible should be emptied into a beaker containing distilled water and stirred with a glass rod; after this the mixture should be filtered and the filtrate evaporated on a watch-glass. It will be found that the substance obtained by heating chalk is slightly soluble. Has the insoluble substance become soluble or has a new soluble substance been obtained by the action of heat on chalk?

As this question cannot be satisfactorily answered now, it must be entered in the question-books.

Powdered iron pyrites should now be heated to a dull

red heat; if a small amount of water be used and the filtrate left to evaporate slowly for a week, green crystals of iron sulphate will be formed; thus a new substance has evidently been obtained, so that heat has changed iron pyrites. These crystals should be dried and kept.

The following method of preparing these crystals given by Glauber in his *Philosophical Furnaces* is of interest:—

" Commonly in all fat soyles or clayie grounds, especially in the White, there is found a kind of stone, round or oval in form, and in bigness like unto a pigeon's or hen's egg, and smaller also, viz. as the joynt of ones finger, on the outside black, and therefore not esteemed when it is found, but cast away as a contemptible stone, which if it be cleansed from the earth, and beaten to pieces lookes within of a fair yellow and in streakes, like a gold marcasite, or a rich gold Oare ; but there is no other taste to be perceived in it, than in another ordinary stone ; and although it be made into a powder and boyled a long time in water, yet it doth not alter at all, nor is there in the water any other taste or color, than that which it had first (when it was poured upon the stone) to be perceived. Now this stone is nothing else, but the best and purest minera (or Oare) of vitrioll, or a seed of metals ; for nature hath framed it round like unto a vegetable seed, and sowed it in the earth, out of which there can be made an excellent medicine as followeth—

Take this oare or minera beaten into pieces, and for some space of time, lay or expose it to the coole aire, and within twenty or thirty days it will magnetically attract a certain saltish moysture out of the aire, and grow heavy by it, and at last it falleth asunder to a black powder, which must remaine further lying there still, untill it grow whitish, and that it do taste sweet upon the tongue like vitriol. Afterward put it in a glass vessel, and poure on so much faire raine water as that it cover it one or two inches ; stirre it about several times in a day, and after a few dayes, the water will be coloured green, which you must poure off, and poure on more faire water, and proceed as before, stirring it often untill that also come to be green ; this must be repeated so often until no water more will be colored by standing upon. Then let all the green waters which you poured off run through filtring paper, for to purifie them, and then in a glass body cut off short let them evaporate till a skin appear at the top : then set it in a cold place, and there will shoote little green stones, which are nothing else but a pure vitriol."

Crystallisation. As the crystals of iron sulphate are the first crystals to be obtained by the class the children will be very eager to find out whether other substances will crystallise, and experiments can be tried with salt, nitre, alum, sal-ammoniac and washing-soda ; and the best specimens of the crystals kept.

Since it has been discovered by the children that the soluble inorganic substances they are acquainted with can be made to crystallise, they will enquire whether the ash obtained by heating the organic substances can also be made to crystallise. This question they can try to answer by treating the beetroot ash with water, heating, filtering it, and then after concentrating the filtrate leaving it to crystallise. When the crystals obtained are examined the children will be surprised to find that they are not all alike, so that their next problem is to try and separate the crystals. This they can do by re-dissolving the crystals, making a hot concentrated solution and allowing it to cool, then pouring the mother liquor off the first formed crystals into a crystallising dish, allowing it to stand and again pouring off the mother liquor from the next formed crystals into another dish and leaving this to crystallise.

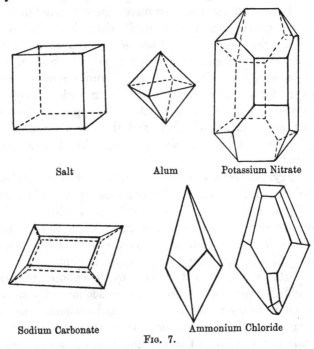

Salt   Alum   Potassium Nitrate

Sodium Carbonate   Ammonium Chloride

Fig. 7.

By redissolving and recrystallising the various products they will find that they can separate the crystals and that they obtain a large proportion of one kind. These crystals they can dry and keep, labelling them "potashes." In this way the class can be introduced to the idea of separation of chemical substances by crystallisation.

A convenient way of obtaining sufficient beetroot ash for the use of the class is to scrape the white sugar-beet on a rough grater, then dry it in the drying oven, after which the desiccated beetroot can be heated to a dull red heat in the muffle furnace until a white ash is obtained.

While waiting for the potashes to crystallise the children should be asked to bring the petals of flowers, especially blue or violet petals, some leaves and "red cabbage." Then they should put into separate watch-glasses some vinegar, lemon juice, ammonia, lime water, a solution of washing-soda, a solution of potashes (liquids the children will all be familiar with), and try the effect of dipping the petals and leaves into the different solutions. They can then divide their solutions into two groups, those which turn the cabbage red and those which turn it green, that is to say into acids and alkalies. They will realise by tasting the solutions that each group has characteristic properties. Litmus can be given to the class as another vegetable substance prepared from a lichen, the so-called "Dyer's Moss." The children will find that by its aid also they can distinguish between acid and alkaline solutions, and they can be told that this fact was first discovered by Robert Boyle.

If time permits the children can prepare their own books of litmus paper.

At this point the effect of heat on the various crystals they have obtained can be tried. Salt they will find does not change in appearance. Washing soda first liquefies, then dries and becomes powdery. Nitre first liquefies and then appears to boil. If a little piece of wood or charcoal be dropped into the crucible it takes fire and burns. Iron sulphate crystals first turn brown and then form a greyish white powder while fumes which are

Fig. 8. Boyle.

strongly acid are given off, and finally a red-brown powder is left. The causes of these changes must next be investigated.

Water of Crystallisation. To do this, weigh an evaporating dish and small funnel, then place some powdered washing-soda in the dish and weigh again; cover the dish with the funnel and heat gently on a sand-bath; drops of moisture will form on the funnel and vapour will escape through the stem. This vapour does not change the colour of litmus paper and if a watch-glass be held over the top of the funnel some of the vapour will condense on it and the liquid formed will be found to be tasteless. A sufficient quantity of this liquid can be collected by attaching a glass tube to the funnel by means of a piece of indiarubber tubing, and allowing the liquid to condense in a test tube, to enable the boiling point and the freezing point of the liquid to be determined and the liquid shown to be water. When dry weigh the dish, substance and funnel again, and find the percentage loss in weight. The difference in weight seems to be due to the loss of water ; can the crystals be reformed by the addition of water ? This can be tried, some of the substance being redissolved and put in a watch-glass to crystallise. The crystals will be once more obtained, so that the effect of heat on washing-soda is to drive off the water of crystallisation from the crystals.

In the case of iron sulphate there is clearly a further change ; after the water of crystallisation has been driven off, on further heating acid white fumes are formed, which will not condense on a watch-glass ; can these white fumes be collected in water ? To do this experiment, first powder the iron sulphate, then heat gently in an evaporating dish on a sand-bath, stirring with a glass rod until the whole is white and powdery ; then transfer the powder to a small glass retort the end of which passes

into a test-tube half full of water, but hardly touches the
surface of the water.   When the powder has all turned

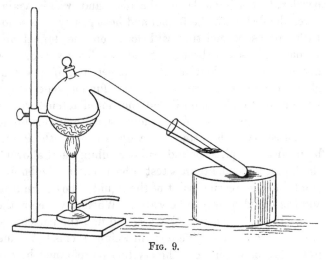

Fig. 9.

red, remove the test-tube and cool the retort slowly.
Some of the red powder can be kept.

Sulphuric Acid.   The liquid in the test-tube is
found to be acid, and on evaporation an oily liquid is
obtained which chars litmus paper, and when heated
strongly, gives off thick white acid fumes.   The old name
for this oily liquid is "Oil of Vitriol," the modern name
"Sulphuric Acid."   The following description of the
preparation of vitriol given by Glauber in the seventeenth
century is of interest :—

"To distill Vitriol there needs no other preparation, but only
that it be well viewed, and if there be any filth amongst it, that
the same be carefully picked out, lest being put together with the
Vitriol into the distilling vessel the spirit be corrupted thereby.

But he that will go yet more exactly to work, may dissolve it in faire water, then filtrate it and then evaporate the water from it

A. The furnace together with the iron diftilling-veffel faftened into it, whereunto a receiver is applyed. B. The diftiller with his left hand taking off the lid, and with his right cafting in his prepared matter. C. The external form of the diftilling veffel. D. The internal form of the veffel. E. Another diftilling veffel, which is not faftened to a furnace, but only ftandeth upon coals.

THE

Fig. 10.

till a skin appear at the top, and then set it in a cold place, and let it shoote again into Vitrioll; and then you are sure that no impurity is left in it. Now your vessel being made ret hot, with an iron ladle cast in one or two ounces of Vitrioll at once, put on the lid, and presently the spirits together with the phlegme will come over into the receiver, like unto a white cloud or mist; which being vanished and the Spirits partly settled carry in more vitriol and continue this so long, until your vessel be full. Then uncover your vessel, and with a pair of tongs or an iron ladle take out the Caput Mortuum and cast more in; and continue this proceeding as long as you please, still emptying the vessel when it is filled, and then casting in more matter and so proceeding until you conceive that you have got spirits enough. Then let the fire go out and let the furnace coole; take off the receiver and poure that which is come over into a retort and lay the retort in sand and by a gentle fire distill the volatile spirit from the heavy oil; having first joyned to the retort the receiver, which is to receive the volatile spirit, with a good lutuum, such as is able to hold such subtle spirits. All the volatile spirit being come over, which you may know by the falling of bigger drops, then take off the receiver and close it well with wax, that the spirit may not make an escape; then apply another (without luting it) and so receive the phlegme by itself, and there will remain in the retort a black and heavy corrosive oyle, which if you please you may rectifie, forcing it over by a strong fire, and then it will be clear."

If the liquid obtained by heating ferrous sulphate and collecting the fumes formed in water be compared with the sulphuric acid in the laboratory, it will be found that the two liquids have the same properties.

If the acid be added slowly to water, heat is generated, also if a weighed quantity be left exposed to the atmosphere it increases in weight. This increase in weight can be shown to be due to the absorption of moisture by placing equal quantities of water in two evaporating dishes, one of these dishes being placed in a desiccator containing sulphuric acid, the other in a desiccator without sulphuric acid; it will be found that the water in the one containing sulphuric acid disappears more quickly than the water in the desiccator containing no acid; so that concentrated sulphuric acid absorbs moisture and can be used as a drying agent. The effect of pouring sulphuric acid on various substances can next be tried, using common substances suggested by the children. Sugar will be seen to turn black and increase in bulk; pieces of cloth to char and drop to pieces, showing the necessity of care in dealing with sulphuric acid.

Hydrochloric Acid. When sulphuric acid is poured on salt there is a brisk effervescence and white acid fumes are given off. The effect of passing these fumes into water can next be tried, as was done in the case of the fumes obtained by heating iron sulphate.

FIG. 11.

For this purpose an apparatus similar to that shown in Fig. 11 is recommended.

The solution obtained is found to be very acid but no oily liquid is produced when it is evaporated, and white fumes are not formed as in the case of sulphuric acid, showing that a new acid has been prepared. On account of its original preparation this acid was formerly called "Spirit of Salt" or "Muriatic Acid"; it is now known as "Hydrochloric Acid"; the reason for which this name has been given to it will be understood by the children later on in the course. This acid, when poured on sugar only turns it yellow; it destroys, but does not char cloth; the concentrated acid fumes in air. If a collecting jar be filled with these fumes and inverted in a trough of water, the water will be seen to rise and fill the jar, showing that the gas is very soluble in water.

When it is found that there is no more effervescence when sulphuric acid is poured on the salt in the flask, the contents of the flask can be put into a beaker of water, and when all the solid has dissolved the solution can be poured into a crystallising dish and left to crystallise.

The crystals formed must be drawn and then kept for future reference.

Extract from Glauber.

How the Spirit of Salt is to be distilled.

"Mix salt, and vitriol or allome together, grinding them very well in a morter (for how much the better they are ground the more spirit they yield). Then cast this mixture into the fire with an iron ladle, viz. as much of it as will be sufficient to cover the coals, and then with a great fire the spirits come forth into the receiver, where being coagulated, they distill down into a dish, and thence into another receiver. And if thou knowest how to work aright, the spirits will like water continually run out through a pipe the thickness of a straw; and thou mayest easily any hour make a pound of the spirit."

Glauber used the furnace shown in the accompanying diagram for this purpose.

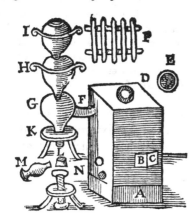

A. The afh hole with the widenefs of the furnace. *B.* the middle hole by which the coals and matter to be diftilled are caft in. *C.* A ftopple of ftone which is to ftop the faid hole after injeétion. *D.* The upper hole with a certain falfe bottome which is to be filled with fand. *E.* The cover of the upper hole, which is laid on after the injeétion of the coales and materials. *F.* A pipe going out of the receiver, and jóyned to the firft pot. *C.* The firft receiver, *H.* The fecond. *I.* The third. *K.* A ftoole on which the firft receiver ftands, having a hole in the middle, through which the neck of the firft pot to which a difh is annexed, paffeth. *L.* The difh, through the pipe whereof the refrigerated fpirits diftill. *M.* A receiver into which the fpirits colleéted in the difh do flow. *N.* A fcrew to be raifed higher at pleafure for the better joyning the receiver to the pipe, and it goeth through a ftoole. *O.* The place of the pipe for the diftilling of fpirit of Vitriol and Allome. *P.* A grate confifting of two ftrong crofs iron bars, faftened in the furnace, and toure or five more lefs that are moveable, for the better cleanfing of the furnace.

G. The

FIG. 12.

Glauber also recommends its use in the kitchen, for as he says :—

"With the help thereof are prepared divers meats for the sick as well as for those that are in health, yea and better than will vinegar, and other acid things ; and it doth more in small quantity than vinegar in a great. Being mixed with sugar it is an excellent sauce for rost meat. It preserves also divers kinds of fruits for many years. Beefe being macerated with it becomes in a few days so tender, as it had been a long time macerated with vinegar. Such and many more things can the spirit of salt do."

Nitric Acid. When sulphuric acid is poured on nitre there is no change till the mixture is heated, and then brown acid fumes are given off which condense and

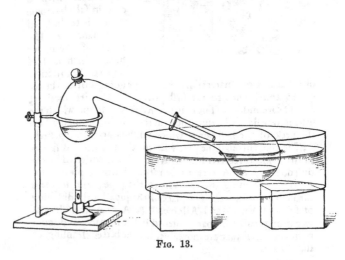

Fig. 13.

form a yellow liquid. This experiment can be most easily done by heating together the nitre and sulphuric acid in a retort, the end of which passes into a flask floating in a trough of water and kept cool by a stream of cold water or by damp filter papers.

The yellow liquid which collects in the flask is very acid and stains the skin yellow; it does not char cloth, but turns it yellow and destroys it. When heated in an evaporating dish in the fume chamber irritating yellow fumes are given off and no residue is left. This acid is called Nitric Acid, the old name being "Aqua Fortis." The residue in the retort must be redissolved and the solution allowed to crystallise. As before, the crystals must be drawn, and then kept for future reference.

Carbonic Acid. When the action of sulphuric acid on washing-soda is tried, a brisk effervescence takes place, just as it did when sulphuric acid was poured on salt; but although the pupils cannot see any fumes in this case, they can prove that something is being given off by fitting the test-tube with a cork through which passes a glass tube bent at right angles leading into another test-tube containing water. They will see bubbles of some gas rising in the water and the liquid in the test-tube will now be acid to litmus. In order to prepare a larger quantity of this acid liquid, a flask should be fitted with a thistle funnel and delivery tube leading into a beaker of water. Some washing-soda is placed in the flask and dilute sulphuric acid poured down the funnel.

When there is no effervescence on the addition of sulphuric acid to the mixture in the flask, the contents of the flask should be poured into a crystallising dish and left to crystallise, and as before, the crystals formed should be drawn and kept. The liquid in the beaker can be examined; it will be found that it turns litmus a port-wine colour and has no effect on cloth. When heated the liquid soon appears to boil, and rapidly becomes less acid, showing that the gas is being driven off.

A similar result is obtained when dilute sulphuric acid is poured upon chalk and upon potashes. This weak acid is termed "Carbonic Acid."

On finding that no new substances are obtained by treating the other inorganic solids they know with sulphuric acid, the children can try the action of alkalies on these solids.

Lime is the commonest alkali they know, but they will find that when they heat it with nitre or common salt in a test-tube no change seems to take place.

Ammonia. When, however, it is heated with sal-ammoniac they will notice a strong smell of ammonia,

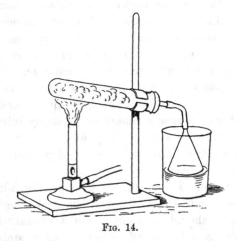

Fig. 14.

which most of the children will at once recognise, and when damp litmus paper is held at the mouth of the test-tube it turns blue, showing that some gas is being given off. This gas can be passed into water. The apparatus most convenient for this experiment is shown in Fig. 14.

The solution obtained will smell like ammonia and will be strongly alkaline.

If a collecting jar filled with this gas be inverted over water, when the cover is removed the water will rush into the jar, showing that the gas is very soluble in water. The solution can be shown to resemble the liquid Ammonia in common use, as if either liquid be heated and the gas given off collected it will be found in both cases to have a characteristic smell, to be very alkaline and very soluble in water. Also on evaporating the solutions no residue is left.

Caustic Soda. When lime water is added to a solution of washing-soda, it becomes milky, showing that some change is produced.

If lime be used in this experiment and the mixture heated and then filtered, the filtrate is found to be strongly alkaline, but on concentration no crystals form. When evaporated to dryness a cream-coloured solid is obtained which does not effervesce on the addition of an acid, showing that this substance is not washing-soda. A concentrated solution is strongly alkaline and destroys filter paper and cloth. This substance is called Caustic Soda. If some of this substance be heated on a platinum wire in a Bunsen flame, the flame is coloured yellow, and washing-soda and salt will be found to produce the same effect, while potashes will give a violet colouration.

Caustic Potash. Potashes can now be heated with

lime as in the case of washing-soda; again a non-crystallisable caustic substance is obtained, which colours the Bunsen flame violet. This substance is called "Caustic Potash."

The names of the acid and alkaline substances known to the class should now be collected and written on the blackboard, and the following list will be obtained :—

Acids.	*Alkalies.*
Sulphuric acid.	Lime Water (Calcium Hydroxide).
Hydrochloric acid.	Ammonia (Ammonium Hydroxide).
Nitric acid.	Caustic Soda (Sodium Hydroxide).
Carbonic acid.	Caustic Potash (Potassium Hydroxide).
	Washing-soda (Sodium Carbonate).
	Potashes (Potassium Carbonate).

CHAPTER II.

Neutralisation. The children now know a certain number of substances which turn red litmus blue and some which turn blue litmus red. They will naturally want to try the effect of a mixture of an acid and an alkali on litmus. The class can be allowed to try the effect of adding hydrochloric acid to a solution of ammonia coloured blue with litmus, using a burette for the purpose; at a certain stage they will get a purple solution, and the term " neutral " will suggest itself as applicable to that which is neither acid nor alkaline. The neutral solution can now be evaporated to dryness. and as the pupils have previously found that both hydrochloric acid and ammonia disappear when heated, they will not expect to find any residue. They can test the white substance that is obtained by heating it in a dry test-tube with a piece of litmus paper held at the mouth of the test-tube, and it will be found that at first alkaline and then acid and finally neutral fumes are given off, while some of the substance sublimes. If some of the solid be dissolved in water and crystals prepared, it will be recognised as sal-ammoniac. To confirm this conclusion these tests can be repeated, using sal-ammoniac from the laboratory.

When concentrated hydrochloric acid and ammonia are brought near each other, white fumes are formed. The cause of this can be discussed, and this property will suggest itself as a means of identifying either hydrochloric acid or ammonia.

The pupils can next try the effect of neutralising caustic soda solution with hydrochloric acid and of evaporating the neutral solution. Once again a white powder is obtained, which on being tasted, crystallised, heated in a test-tube and on a platinum wire is found to be common salt. The class can now be told that by chemists sal-ammoniac is also called a salt, and that this term is applied to all substances obtained by neutralising an acid with an alkali.

From this point careful measurements should be taken, and solutions of acids and alkalies of known strength used.

As the children at this stage cannot themselves prepare solutions of the acids or of ammonia of a known strength, this had better be done by the teacher, and semi-normal solutions are recommended for this purpose.

The children can prepare for themselves solutions of caustic soda, sodium carbonate, caustic potash and potassium carbonate. Caustic soda solution should contain 2 gms. of caustic soda per 100 c.c. of solution. Sodium carbonate solution should contain 2·65 gms. of anhydrous sodium carbonate per 100 c.c. of solution. Caustic potash solution should contain 2·8 gms. of caustic potash per 100 c.c. of solution. Potassium carbonate solution should contain 3·45 gms. of anhydrous potassium carbonate per 100 c.c. In carrying out the next experiments each group of pupils should be provided with two burettes fixed in a stand, having the burettes labelled "acid" and "alkali" respectively, also with an evaporating dish covered by a watch-glass, which should be weighed at the beginning of every experiment.

A measured quantity of the acid is then run into the evaporating dish, a drop of litmus is added to act as an indicator, and the alkali run in slowly from the burette till the solution is neutral. If time permits a second experiment should be tried without litmus, using the same quantities of the acid and of the alkali as were used in the previous experiment.

The evaporating dish should then be covered with a watch-glass and the solution put to evaporate either on a sand-bath or on a water-bath according to the substance formed. After the solution has been evaporated to dryness the dish and watch-glass should be reweighed and the amount of salt formed calculated.

If the same quantity of acid be used in every case, the children will be able to compare results more easily than if varying quantities were used, which would necessitate calculations difficult for children of this age.

To make this work valuable to the class the teacher is advised to select the experiments suggested by the children in the following order :—

A. 25 c.c. of Hydrochloric Acid neutralised with :—

    1. Caustic Soda solution.
    2. Caustic Potash solution.
    3. Ammonia solution.
    4. Lime Water.
    5. Sodium Carbonate solution.
    6. Potassium Carbonate solution.

B.   25 c.c. of Nitric Acid neutralised with :—

       1.   Caustic Potash solution.

       2.   Caustic Soda solution.

       3.   Ammonia solution.

       4.   Lime Water.

       5.   Sodium Carbonate solution.

       6.   Potassium Carbonate solution.

C.   25 c.c. of Sulphuric Acid neutralised with :—

       1.   Caustic Soda solution.

       2.   Caustic Potash solution.

       3.   Ammonia solution.

       4.   Lime Water.

       5.   Sodium Carbonate solution.

       6.   Potassium Carbonate solution.

The results of these experiments should lead to the preparation by the children of twelve different salts and to the discovery of their more important properties; at the same time the practical uses of these substances in everyday life or in manufactures should be brought before the children and any facts of historical interest about the substances told them. The following particulars concerning these salts may be found useful :—

A 1. *Sodium Chloride produced by neutralisation of Hydrochloric Acid with Caustic Soda.*

A white solid which crystallises in cubes, is readily soluble in water and colours the Bunsen flame yellow.

The crystals do not contain any water of crystallisation. Sodium chloride is used for culinary purposes and in the manufacture of soda. On an average 2,192,965 tons are used annually in England.

Specimens of Rock-salt ought to be shown to the

children and the cleavage of the crystals pointed out, while for home-work they could be told to find out all that they can about salt-mines and the preparation of salt from sea-water.

At the next lesson the facts found out by the children can be discussed and a photograph of a salt-mine shown to them.

A 2. *Potassium Chloride produced by neutralisation of Hydrochloric Acid with Caustic Potash.*

This substance resembles sodium chloride in its crystalline form, but the taste is more bitter and it colours the Bunsen flame purple.

It is obtained in large quantities from the salt beds of Stassfurt, and the children should be encouraged to find out as much as possible about these deposits.

This salt is used in the manufacture of nitre and is employed as a fertiliser for artificial manures.

A 3. *Ammonium Chloride produced by the neutralisation of Hydrochloric Acid with Ammonium Hydrate.*

This salt will be easily recognised by the class by the way in which it sublimes when heated, and by alkaline and then acid fumes being given off. It crystallises in

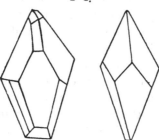

Fig. 15.   Ammonium Chloride.

arborescent or feather-like growths which consist of an
aggregation of small octahedrons.

This substance was brought from Asia to Europe
in the seventh century and was probably derived from
the volcanoes of Central Asia. Its original name was
changed to Sal-ammoniacum, which was the name which
had been given first to the common salt found in the
neighbourhood of the ruins of the temple of Jupiter
Ammon; then, as in later times, Sal-ammoniac was
prepared in Egypt from the salt obtained by burning
camel's dung.

Ammonium chloride is used in electric batteries; it
is also a valuable medicine, and is used in dyeing, and
in the printing of calico.

A 4. *Calcium Chloride produced by the neutralisation
of Hydrochloric Acid with Lime Water.*

The strength of the lime water used in the production of this salt can be estimated by evaporating
50 c.c. of lime water to dryness, igniting the residue and
weighing it.

A characteristic property of calcium chloride is the
rapidity with which it absorbs moisture, and, on account
of this property, it is very difficult to crystallise it, but if
a concentrated solution be left in a desiccator crystals
can be obtained. It colours the Bunsen flame brick-red.

The salt was known in the fourteenth century by
the name " Sal-ammoniacum fixum," and was originally
prepared by heating together sal-ammoniac and lime.

It is used in the laboratory as a drying agent, being
rather more convenient for ordinary use than concentrated sulphuric acid.

A 5. *Sodium Chloride produced by the neutralisation of Hydrochloric Acid with Sodium Carbonate.*

The children will find that when the acid has been neutralised by the sodium carbonate solution and the liquid boiled, it becomes alkaline. A little thought will make them realise that some of the carbonic acid gas had been dissolved in the water, and when the solution was heated it was driven off. They will also soon suggest that the simplest way of obtaining a neutral solution is to add the acid to a hot solution of the alkali, boiling the solution at intervals to drive off any dissolved carbonic acid gas.

At the end of the experiment they must calculate the amount of sodium carbonate solution necessary to neutralise 25 c.c. of the acid. They will then find that sodium chloride is again produced and that the amount of it is the same as in the previous experiment.

A 6. *Potassium Chloride produced by the neutralisation of Hydrochloric Acid with Potassium Carbonate solution.*

In this experiment it will be found that the same precautions must be taken as in the previous experiment with sodium carbonate. Potassium chloride is again formed and the same amount of salt produced as when hydrochloric acid is neutralised with caustic potash.

All the results obtained by neutralising 25 c.c. of hydrochloric acid solution can now be collected by the teacher and written on the blackboard. It will be seen that a different weight of each alkali is required to neutralise the same amount of the acid, but that in the case of caustic soda and sodium carbonate, although

different weights of the alkalies are used the same
weight of sodium chloride is produced; similarly with
caustic potash and potassium carbonate the same weight
of potassium chloride is produced.

It will be found in many cases that the children will
themselves suggest writing out the results in the form of
equations, but it must be clearly understood that symbols
cannot be introduced at this stage, the equations being
merely written in the following way :—

Hydrochloric Acid + Caustic Soda solution

　　= Sodium Chloride + Water.

Hydrochloric Acid + Sodium Carbonate solution

　　= Sodium Chloride + Carbonic Acid gas + Water.

Hydrochloric Acid + Caustic Potash solution

　　= Potassium Chloride + Water.

Hydrochloric Acid + Potassium Carbonate solution

　　= Potassium Chloride + Carbonic Acid gas + Water.

Hydrochloric Acid + Ammonia solution

　　= Ammonium Chloride + Water.

Hydrochloric Acid + Lime Water

　　= Calcium Chloride + Water.

The weights of the substances used and the salts
obtained should be entered under the names, and in the
case of the hydrates, it can be readily proved that these
statements are true quantitatively as well as qualitatively.
This can be done by finding the weight of 1 c.c. of the
acid solution and of 1 c.c. of the alkali solution; as
the weight of hydrochloric acid gas and also the weight of
alkali in each c.c. of the solution is known, the weight of
water is easily calculated; if the neutral solution be

weighed before and after evaporation it will be seen that the amount of water has increased during the reaction. The liquid formed can be shewn to be water by means of the tests used in previous experiments.

B 1. *Potassium Nitrate produced by the neutralisation of Nitric Acid with Caustic Potash solution.*

When some of this salt has been crystallised and the crystals compared with the crystals known to the children, it will be recognised as potassium nitrate : this can be confirmed by tasting some of it, heating some on a platinum wire and some in a dry test-tube, and then

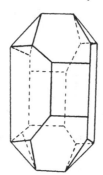

Fig. 16.   Potassium Nitrate.

dropping a piece of charcoal into the fused substance. The crystals are long six-sided prisms with wedge-shaped ends.   Fig. 16.

This substance was known to the ancients.   In some hot countries such as India, after the rainy season, the soil becomes covered with small crystals of nitre.   It is now manufactured from Chili salt-petre and potassium chloride.

Large quantities of this substance are used in the manufacture of gunpowder and other explosives; for making fireworks and lucifer matches, and for the salting or pickling of meat.

B 2. *Sodium Nitrate produced by the neutralisation of Nitric Acid with Caustic Soda Solution.*

This substance will be found to crystallise in obtuse rhombohedrons and the crystals will absorb moisture rapidly from the air.

When heated in a dry test-tube it behaves in a similar manner to that of potassium nitrate, but when heated on

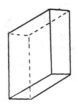

Fig. 17.   Sodium Nitrate.

a platinum wire the flame colouration is yellow, instead of violet.

Sodium nitrate is largely used in the manufacture of nitric acid and of potassium nitrate, large quantities of the salt being brought for this purpose from Peru.

B 3. *Ammonium Nitrate produced by the neutralisation of Nitric Acid with Ammonia solution.*

The neutral solution must be evaporated on a water-bath.

When some of the salt is redissolved and left to crystallise, beautiful characteristic crystals are obtained in the form of rhombic prisms.

When the crystals are heated in a dry test-tube, at first they fuse and seem to boil, then after an interval there is a slight explosion, and a tongue of flame is seen in the test-tube, followed by brown fumes. The teacher must warn the children to use only a small quantity of the salt. When ammonium nitrate is dissolved in water the temperature is lowered. The class will find on examination that the salt labelled ammonium nitrate in the laboratory will have the same properties. It must be entered in the question books that ammonium nitrate is

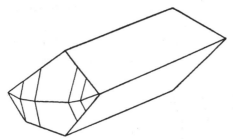

FIG. 18.   Ammonium Nitrate.

a substance to be examined in more detail later on, and its uses will then be better understood. This salt was first prepared by Glauber.

B 4.   *Calcium Nitrate produced by the neutralisation of Nitric Acid with Lime Water.*

This salt forms a white porous mass, which rapidly absorbs moisture from the air; when some of the substance is dissolved and the solution left to crystallise in a desiccator prismatic crystals are formed. It can be distinguished from calcium chloride by its behaviour when heated in a dry test-tube, and by dropping a piece of charcoal into the fused salt. It gives the Bunsen flame a brick-red colouration.

This salt was first prepared in 1674 by the alchemist Baldewein, who noticed that this substance, when heated and then exposed to sunshine, appeared luminous in the dark. From that time the substance so prepared has been known as " Baldwin's Phosphorus." This substance is sometimes used in the laboratory as a drying agent, and occasionally for the artificial preparation of nitre.

B 5. *Sodium Nitrate produced by the neutralisation of Nitric Acid with Sodium Carbonate solution.*

When carrying out this experiment the class will realise that the same precautions must be taken as when hydrochloric acid was neutralised with sodium carbonate solution.

B 6. *Potassium Nitrate produced by the neutralisation of Nitric Acid with Potassium Carbonate solution.*

In this experiment also it will be found necessary to neutralise the carbonate with the acid and to calculate the amount of carbonate necessary to neutralise 25 c.c. of the acid.

From the neutralisation of Nitric Acid with the various alkalies the following results are obtained:—

Nitric Acid + Caustic Potash solution

    = Potassium Nitrate + Water.

Nitric Acid + Potassium Carbonate solution

    = Potassium Nitrate + Water + Carbonic Acid gas.

Nitric Acid + Caustic Soda solution

    = Sodium Nitrate + Water.

Nitric Acid + Sodium Carbonate solution

    = Sodium Nitrate + Water + Carbonic Acid gas.

Nitric Acid + Ammonia solution

    = Ammonium Nitrate + Water.

Nitric Acid + Lime Water

    = Calcium Nitrate + Water.

As in the case of the chlorides, the weights of the substances used and the salts obtained should be written under the equations.

Again it will be evident to the children that different weights of the various alkalies are required to neutralise the same amount of acid, and that the same weight of sodium nitrate is produced by neutralising 25 c.c. of nitric acid with caustic soda or sodium carbonate, although different weights of these alkalies are used.

C 1. *Sodium Sulphate produced by the neutralisation of Sulphuric Acid with Caustic Soda solution.*

This substance will crystallise out from its solution in colourless monoclinic prisms, which effloresce on exposure

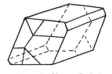

FIG. 19.   Sodium Sulphate.

to dry air. It readily forms a supersaturated solution, which crystallises suddenly with a considerable rise of temperature when a small crystal is thrown into the liquid.

This salt is first mentioned in Glauber's work, *De Natura Salium*, in 1658. Glauber obtained it from the residue left in the preparation of hydrochloric acid, and believed it to be possessed of valuable medicinal properties. It is found in large quantities in certain mineral springs such as those at Friedrichshall, and is obtained in large quantities in the manufacture of washing-soda. The commercial name of sodium sulphate is "salt-cake," while its popular name is "Glauber's salt."

C 2. *Potassium Sulphate obtained by the neutralisation of Sulphuric Acid with Caustic Potash solution.*

When some of this salt is dissolved in water and left to crystallise, small hard rhombic pyramids will form.

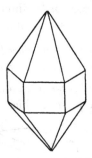

Fig. 20.   Potassium Sulphate.

This salt can be recognised by the children by its bitter saline taste, by the violet colouration it gives to the Bunsen flame, and by the shape of the crystals.

Potassium sulphate was known to Glauber and Boyle. It is used medicinally and in the manufacture of potash, alum, and potassium carbonate.

C 3. *Ammonium Sulphate produced by the neutralisation of Sulphuric Acid with Ammonia solution.*

When some of this salt is allowed to crystallise, the crystals are found to be isomorphous with those of potassium sulphate; this will give the children their first idea of isomorphism.

They will find that the properties of ammonium sulphate are not the same as those of potassium sulphate, as when heated in a dry test-tube ammonia fumes are given off; also it will be found not to colour the Bunsen flame purple.

It is used in the preparation of other ammonia salts and largely employed as a fertiliser in artificial manures. Glauber is again said to be the first chemist who accurately described ammonium sulphate.

C 4. *Calcium Sulphate produced by the neutralisation of Sulphuric Acid with Lime Water.*

After evaporating the neutral solution to dryness, it will be found that, if a little water is added to the solid obtained, the mixture will become hot and in a short time will set into a hard mass. From this characteristic property it will be easily identified by the class as " Plaster of Paris." This substance is found in the natural state as the mineral "gypsum," and when in large monoclinic crystals is known as selenite.

Calcium Sulphate is only very slightly soluble in water, but when present causes the " permanent hardness " of water.

It is largely used as a cement, for ornamental plaster work, and for making plaster casts; it is also used by paper makers as a filling for writing paper.

C 5. *Sodium Sulphate produced by the neutralisation of Sulphuric Acid with Sodium Carbonate solution.*

C 6. *Potassium Sulphate produced by the neutralisation of Sulphuric Acid with Potassium Carbonate solution.*

When preparing these salts the same precautions must be taken as in the previous experiments with carbonates.

The results obtained by neutralising Sulphuric Acid with the various alkalies can now be collected and expressed in the form of equations:—

Sulphuric Acid + Caustic Soda solution

    = Sodium Sulphate + Water.

Sulphuric Acid + Sodium Carbonate solution

    = Sodium Sulphate + Water + Carbonic Acid gas.

Sulphuric Acid + Caustic Potash solution

    = Potassium Sulphate + Water.

Sulphuric Acid + Potassium Carbonate solution

    = Potassium Sulphate + Water + Carbonic Acid gas.

Sulphuric Acid + Ammonia solution

    = Ammonium Sulphate + Water.

Sulphuric Acid + Lime Water

    = Calcium Sulphate + Water.

The same general conclusions will be arrived at by the class from the results obtained by the neutralisation of Sulphuric Acid as from the results obtained by the neutralisation of the other acids.

Carbonic Acid has not yet been neutralised by these alkalies, and now experiments can be tried to find out whether, when caustic soda is neutralised by this acid, the same salt is produced as when sodium carbonate is neutralised by it. As carbonic acid solution is only feebly acid the gas itself had better be used for this purpose. The gas can be easily obtained by the action of dilute acid on lumps of washing-soda. In one small flask put 25 c.c. of a concentrated solution of caustic soda, and in another 25 c.c. of a concentrated solution of sodium carbonate. Colour these solutions with litmus and pass carbonic acid gas into them for some time. The colour of the litmus will be found not to change ; the children will conclude therefore that the carbonic acid gas has not combined with these alkalies. In order to test this

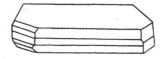 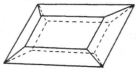

Sodium bicarbonate.   Fig. 21.   Washing-soda crystal.

conclusion, pour the solutions into crystallising dishes and leave them to crystallise.

It will be discovered that the same new substance is formed in both cases, the crystals of which are unlike those of sodium carbonate, and do not effloresce when exposed to air in the way sodium carbonate crystals do.

These crystals should be kept for reference. In order to find out more about this new substance, the experiment had better be repeated, only this time, instead of the solutions being left to crystallise, they must be heated, and the children will at once notice there is an evolution of gas, so that they all expect to find caustic soda and sodium carbonate respectively again produced when the solutions are evaporated to dryness. But on examination the residues will be found to be alike. The crystals formed from both residues resemble those of sodium carbonate, effloresce on exposure to air and effervesce on the addition of an acid; while from other tests, such as alkalinity and the flame colouration, it will be discovered that in both cases sodium carbonate has been obtained.

Sodium Bicarbonate. Thus sodium carbonate can be formed by the combination of carbonic acid gas with caustic soda, but on the addition of more carbonic acid gas the new substance is obtained which can now be named " Sodium Bicarbonate." The meaning of the name of course is not explained, but the children will all be familiar with the substance sodium bicarbonate and with its uses in cooking, which they will now be able to understand, especially if they add some sodium bicarbonate to hot water and notice the evolution of gas. Some sodium bicarbonate can be dissolved in cold water and left to crystallise and the crystals formed compared with those obtained from the solution of caustic soda through which carbonic acid gas was passed for some time.

The children will conclude from these experiments that carbonic acid gas can combine with caustic soda in two proportions, producing substances which have distinct properties, but which both effervesce on the addition of an acid.

As a useful exercise the same experiments can be repeated, using caustic potash and potassium carbonate.

The effect of passing carbonic acid gas through lime water can next be tried. Immediately a white precipitate is formed, but after the gas has been passed through the solution for some time it becomes clear again. On

heating the clear solution, bubbles of gas are given off
and the solution again becomes milky ; on evaporating to
dryness, a white powder is obtained which can be shown
to be insoluble in water and therefore is not lime. It
will be realised by the class that this formation of a
precipitate with lime water can be used as a means for
identifying carbonic acid gas. These changes will at first
seem very puzzling to the children, but on comparison
with the previous experiments with caustic soda they will
realise that probably here also two compounds have been
formed, one of which is insoluble and the other soluble in
water ; that is to say, when carbonic acid gas is passed
into lime water the insoluble compound is formed and a
precipitate is obtained ; on continuing to pass the gas
through the mixture the new soluble compound is formed
and the solution becomes clear. On heating the clear
solution and driving off the excess of gas, the insoluble
compound is again obtained. On adding hydrochloric
acid to this insoluble substance, there is a brisk
effervescence and a soluble substance is obtained ; this is
an example of *chemical solution*. On evaporating this
solution to dryness and leaving some of the solid obtained
in a watch-glass exposed to the air, it will be found that
it absorbs moisture rapidly : also if some is heated on a
platinum wire in a Bunsen flame a red colouration will be
observed. The class will recognise this substance as
calcium chloride, which before was obtained by the
neutralisation of hydrochloric acid with lime water.

Lime, they know, is obtained by heating chalk : can
this new substance be chalk ? To answer this question
some of this substance must be strongly heated in a crucible
in a table furnace ; they will find that lime is produced.

Calcium Carbonate. Therefore when carbonic acid gas is passed into lime water, chalk or calcium carbonate is formed.

The weight of carbonic acid gas which combines with lime to form calcium carbonate can next be found by heating in a muffle furnace or table furnace a known weight of chalk and finding the loss in weight. This will be found to be 44 per cent.

The effect of heating a known weight of finely powdered marble in a muffle furnace can also be tried and the loss per cent. calculated: this will be found to be the same as in the case of chalk. On examination the substance left in the crucible is found to be lime, so that marble is probably also calcium carbonate.

By pouring dilute hydrochloric acid on marble and showing that the gas given off turns lime water milky and is therefore carbonic acid gas, while the residue in the beaker is calcium chloride, this view can be confirmed.

Similarly limestone can be shown also to be calcium carbonate. The specimens of chalk, limestone, and marble can once again be examined by the children and now that they know that the chemical composition of these substances is the same, their origin can be discussed, and the chalk particles examined under the microscope.

On referring to the question books, it will be found that one of the questions is, "Why does sodium carbonate solution become milky when lime water is added to it?" This experiment can be tried again and the precipitate on examination will be found to be chalk, while the filtrate is caustic soda, showing that sodium carbonate becomes caustic when deprived of its carbonic acid gas.

This was first proved by Black in the eighteenth century.

Black. This chemist did much research work on

FIG. 22. Black.

carbonic acid and its compounds with the alkalies and the alkaline earths. Before his time the carbonates were regarded as simple substances, but he showed that when limestone was burnt carbonic acid gas was given off, and to this gas he gave the name " fixed air."

Black also showed that quicklime absorbs carbonic acid gas from the atmosphere; this fact will enable the children to understand the hardening of mortar when exposed to air, and they can prove by pouring some acid on old mortar that calcium carbonate has been formed in it.

It will interest the children and make them remember this part of their work if the teacher reads to them extracts from the Alembic Reprint of Black's Researches which bear upon the experiments they have themselves tried. In the same reprint there is an account of Black's researches on Magnesia Alba, a substance which he obtained by heating together potashes and Epsom salt, a substance which the children will probably know. They will find that Magnesia Alba resembles chalk in that it effervesces on the addition of an acid and that it loses weight when heated, leaving a substance in the crucible which in many ways resembles lime.

The children have previously found out that chalk loses 44 per cent. of its weight when heated; they can now try to find out what weight of carbon dioxide is given off when a known weight of chalk is acted upon by hydrochloric acid.

This experiment is easily done if a small flask be weighed and about ·5 gram of powdered calcium carbonate put in the flask, which is reweighed. The flask is now fitted with an indiarubber cork having two holes through which pass a tube bent at right angles and a drying tube containing calcium chloride respectively. The right-

angled tube must be closed by a piece of indiarubber
tubing fitted with a clamp. The cork is now removed
and two small test-tubes containing concentrated hydrochloric acid are suspended by means of fine cotton in the
flask, the chalk having been previously covered with
distilled water. The cork is then refitted, care being
taken to keep the tubes containing the acid upright.
The whole apparatus is then reweighed, after which the
acid in the test-tubes is upset on the chalk. When all
the chalk has dissolved the clamp is removed from the

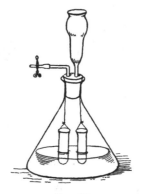

Fig. 23.

piece of indiarubber tubing and air drawn through the
apparatus by means of an aspirator. The clamp is then
replaced and the whole apparatus reweighed. The
difference between the second and third weighings gives
the weight of carbonic acid gas given off, and from this
the percentage loss of weight can be calculated. This
should be 44 per cent.

The class can next try to find the weight of a litre of

carbonic acid gas. This can be done by fitting up an apparatus similar to that shown in Fig. 24.

The calcium carbonate is placed in the flask with a little distilled water and the test-tubes are filled with concentrated hydrochloric acid and suspended in the flask, which together with the drying tube is then weighed. The drying tube is then connected to the Winchester, the acid is upset over the chalk and the clamp connecting the Winchester with the acid tube is at once loosened.

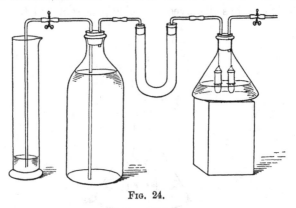

Fig. 24.

When all the chalk has dissolved the level of the water in the Winchester and the measuring jar is adjusted and the clamp between them closed. The drying tube and flask are then disconnected from the Winchester and air is drawn through them by means of an aspirator, after which the apparatus is reweighed. The difference in weight gives the weight of the carbon dioxide while the volume of water in the measuring jar gives the volume of carbon dioxide given off. From these results the weight of a litre of carbon dioxide can be calculated.

CHAPTER III.

The Discovery of Oxygen. Since chalk, marble and limestone lose in weight when heated, the children should find out whether the common metals which they brought at the same time as these building materials also change in weight when heated.

Known weights of lead, copper and zinc should be heated in crucibles over a furnace burner and the changes in appearance carefully noted. After heating for some time the crucibles should be reweighed, and it will be found that in every case there is an increase in weight: this the children will find very difficult to account for and they will want to try experiments to get some solution of the problem. Each metal can now be heated in an iron spoon—the calx as it is formed being removed with an iron rod. Fine copper filings should be used for the purpose, and the pupils will find that the copper turns black. When granulated zinc is heated it becomes yellow while hot, and turns white when cold; it may take fire and burn with a bright white flame.

Lead first melts and then a greenish yellow calx forms. If this be heated for some time and constantly stirred, part of it will appear red while hot, but on cooling will turn yellow again; but if some of this yellow powder be placed on an iron sand tray over a rose burner and kept at a temperature of about 210°C. for 24 hours, taking care to stir it at intervals, it will be found that a red substance has been formed. The red powder is the so-called red lead, while the first formed yellow calx is known as "Litharge." New substances have clearly been formed by heating these metals in air; what will be the effect of treating these new substances with acids? Hydrochloric acid can be tried in each case.

The black substance obtained by heating copper dissolves in the acid, forming a green solution from which, if left to crystallise, green rhombic crystals will be deposited, especially if the operation has been completed in a desiccator.

The zinc white when heated in hydrochloric acid dissolves: on evaporation to dryness a white deliquescent substance is obtained. If, instead of evaporating to dryness, a little concentrated hydrochloric acid is added when the solution becomes syrupy, small octohedral crystals will separate out.

When litharge is treated with hydrochloric acid after heating, as the solution cools, small white crystals are deposited on the top of the undissolved litharge. When heated again these are found to disappear, but if the hot solution be decanted into a watch-glass the white needle-shaped crystals form at once, and the children will find that the whole of the litharge will dissolve and form a solution from which these crystals separate out on cooling. When red lead is treated with concentrated hydrochloric acid and warmed, a yellow stifling gas is given off and a yellow solution is obtained from which the same white needle-shaped crystals separate out.

It is advisable to carry out in the fume chamber the experiments which the children will now try in order to find out the properties of this greenish yellow gas.

When damp litmus paper is held in the gas it is bleached. If the gas is passed into water it will be found that the solution also has bleaching properties, a strong suffocating smell and a yellow colour, but when the solution is left exposed to sunlight the colour and smell disappear, and the children will find that instead of bleaching litmus it now turns blue litmus red.

On evaporating this acid solution there is no residue, and acid fumes are given off, which when brought near ammonia gas form dense white fumes: so this liquid will be identified as a solution of hydrochloric acid gas, and therefore there is evidently some connection between hydrochloric acid gas and this greenish gas; this must be a subject of future investigation.

Chlorine. This greenish gas was first discovered in 1774 by Scheele by treating a mineral ore called pyrolusite with hydrochloric acid. This gas was called by Davy "Chlorine" and was proved by him to be an element; his methods of proof must be discussed later on.

Since Chlorine is in some way connected with hydrochloric acid, and this acid reacts upon an alkali to form a chloride, the question arises, will chlorine when passed into a solution of an alkali in the same way produce a chloride?

The class can try the effect of passing chlorine into a hot concentrated solution of caustic potash. As red lead is rather expensive for class use the children might be allowed to try Scheele's method of preparing this gas and use pyrolusite and hydrochloric acid. To carry out this experiment fit a flask with a two-holed cork through which pass a thistle funnel and a delivery tube bent twice at right angles, the end of the delivery tube dipping into a beaker containing the hot potash solution. About 50 c.c. of concentrated hydrochloric acid should be put into a flask and 20 gms. of powdered pyrolusite added.

On heating the flask chlorine will be evolved and will pass into the caustic potash solution. When the solution is saturated, boil to drive off the excess of chlorine and leave to cool in a crystallising dish. Large plate-like crystals

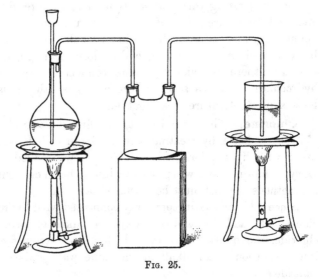

Fig. 25.

of potassium chlorate will form: pour off the mother liquor from these and evaporate it to about one-half and again leave to crystallise : cubical crystals will now form, which can be easily recognised by the children as potassium chloride crystals.

The first formed crystals when dry, if large, will appear iridescent. These heated in a hard glass test-tube crackle and fuse, and then the liquid seems to boil;

this may be due to the evolution of a gas; but if so, it has no effect upon litmus; a glowing splint brought near the mouth of the test-tube at once relights, so that something appears to be given off. The children will want to try, as before, the effect of passing this gas (if it is a gas) into water. On doing this they will see that bubbles of gas rise through the water, but that the water does not seem to change. If a piece of indiarubber tubing be fixed to the end of the delivery tube under the water and a test-tube full of water be inverted over the end, the water will be displaced by a colourless gas, and

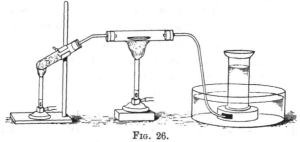

FIG. 26.

if a glowing splint be put into the test-tube of gas, it will burst into flame. The children will naturally try the effect of passing this gas into acid and alkali solutions; on evaporating these solutions they will find that they do not obtain a new substance, so that the new gas does not react with the acids and alkalies they know. As a splint burns so readily in this gas the next step in the investigation will be to try the effect of passing the gas over heated solids.

The class can first try passing this new gas over heated powdered charcoal, using apparatus shown in Fig. 26.

The charcoal will glow brightly and bubbles of gas will

rise through the water. After allowing this to go on for a few minutes, the collecting jar full of water should be inverted over the end of the delivery tube. When the jar is full of gas, cover with a glass plate, and test to find out whether the gas will relight a glowing splint. It will be found that the splint does not burn and that even a lighted taper will be extinguished.

Oxygen. On trying the action of this gas on litmus and on lime water the children will find that the gas is carbonic acid gas. If care be taken to expel all air from the apparatus before the charcoal is heated the children will conclude that carbonic acid gas is a compound of carbon and the new gas, which can now have the name of "Oxygen" or "acid producer" given to it.

A valuable experiment at this point will be to find the percentage composition of carbonic acid gas. This can be easily done in the following way :—

Composition of Carbonic Acid Gas. A combustion tube is weighed; some powdered charcoal put in it and the whole reweighed; then two small flasks containing a concentrated solution of caustic potash are weighed, and the whole apparatus fitted up as shown in Fig. 27.

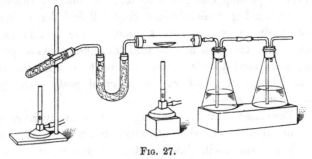

Fig. 27.

The oxygen is prepared by heating potassium chlorate, which is placed in the test-tube and the gas is dried by passing it through the calcium chloride in the U-tube. When all the air has been driven out of the apparatus, heat the combustion tube containing the charcoal and continue heating till nearly all the carbon has been burnt away. Then remove the burner; but continue the flow of oxygen until the combustion tube is cool. Then reweigh the combustion tube to find the weight of carbon used, and reweigh the flasks containing the caustic potash solution to find the weight of the carbonic acid gas which has been formed. From these results the percentage composition of carbonic acid gas can be calculated.

Oxygen can now be passed over heated copper by simply using the apparatus shown in Fig. 26, substituting copper for carbon.

The copper will blacken and the children will recognise the product as appearing similar to that obtained when copper is heated in air. As copper increased in weight when heated in air, the children will want to find out whether copper increases in weight when heated in oxygen and whether the increase in weight is the same, showing that the substances are identical.

A small quantity of finely divided copper must be put in a weighed combustion tube and the whole weighed again to find the weight of the copper. After heating in a current of oxygen until all the copper has turned black, cool and weigh it again, and calculate the percentage increase in weight.

In the next experiment weigh the finely divided copper in a combustion tube as before and connect the end of the combustion tube with an aspirator as shown in Fig. 28.

Heat the copper strongly, and then let the air pass over it slowly. When all the copper has changed colour reweigh and calculate the percentage composition of the black substance. It will be found to be the same as when the copper was heated in oxygen.

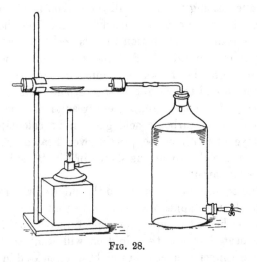

Fig. 28.

Nitrogen. From this experiment it will be concluded that air contains oxygen; but when substances are burnt in air they do not burn as vigorously as they do in oxygen; what is the difference between air and oxygen? On referring to the previous experiment the children will see that some of the air has passed over into the aspirator. Is that residual air the same as ordinary air? When a lighted taper is plunged into the gas in the aspirator it at once goes out; the children will think that they have again obtained carbonic acid gas, but when they pour in lime water and shake it up with the gas in the aspirator the lime water only turns slightly milky; they will on consideration realise that all the gas in the aspirator cannot be carbonic acid gas, and that the comparatively small amount that is present may have come from the air in the room ; so they can try the experiment again, first freeing the air of carbonic acid gas by passing it through a flask containing a concentrated solution of caustic potash. After this has been done, they will find that the gas contained in the aspirator, although it puts out a lighted taper, does not turn lime water milky, and there-fore is not carbonic acid gas, but evidently a new gas. The class can be told that this residual gas until 1894 was regarded as consisting of the element nitrogen, but the researches of Sir William Ramsey and Lord Rayleigh have shown that it is a mixture of several gases, but that nitrogen is the principal constituent. The further in-vestigation of nitrogen must be left for a later stage.

Composition of Air. Air therefore mainly consists of oxygen and nitrogen. To find the relative proportion

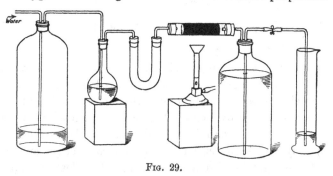

Fig. 29.

of oxygen and nitrogen present in the air, the following experiment can be tried and will probably be suggested by the class.

Connect the combustion tube containing a roll of copper gauze with a Winchester full of water on one side and a U-tube containing small lumps of caustic potash on the other side. The U-tube must be connected with a flask containing concentrated sulphuric acid, which in its turn is connected with another Winchester into which water can be slowly passed by connecting it with a water-tap as shown in Fig. 29. A long delivery tube must lead from the other Winchester to a graduated cylinder. Heat the copper first strongly and allow the water which is expelled on account of the expansion of the air in the combustion tube to flow away: then turn on the water-tap and put the graduated cylinder under the delivery tube. When the graduated cylinder contains 200 c.c. of water turn off the tap and so stop the flow of air through the apparatus, and measure the water which has passed into the first Winchester. This gives the volume of air which contains 200 c.c. of nitrogen and so the composition of air by volume is arrived at.

Density of Air. The density of air can be found in the following way :—

A flask should be fitted with a one-holed cork through which passes a glass tube about 10 cms. long, on the end of which a piece of indiarubber tubing is fitted. and the opening closed by a clip. The flask is then weighed and afterwards the clip is opened and the flask is heated gently over a Bunsen burner to expel some of the air: the clip is again closed and when cool the flask is again weighed; the difference between these weighings will give the weight of the expelled air. The flask is then inverted and the end of the indiarubber tubing put under water and the clip opened: the volume of the water which enters the flask is measured, and thus the weight of a known volume of air is determined.

The children already know from their experiments with sulphuric acid that air contains moisture: they also know it contains carbonic acid gas, and the amount of carbonic acid gas and of water vapour present in the air can easily be estimated by passing air through weighed flasks, the first and last containing concentrated sulphuric acid, and the other two a concentrated solution of caustic potash. These flasks are connected, as shown in Fig. 30.

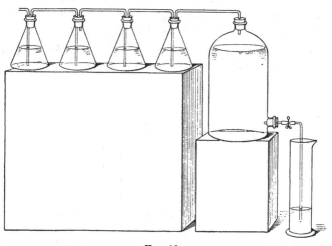

Fig. 30.

The water from the aspirator runs into a measuring cylinder.

If the flasks are reweighed at the end of the experiment the gain in weight of the first flask containing sulphuric acid gives the amount of moisture present, and the gain in weight of the other three flasks gives the amount of carbonic acid gas present: the volume of the water in the graduated cylinder will give the volume of air passed through the flasks.

So far only two oxides have been considered, an oxide of carbon and an oxide of copper. It will be interesting at this stage to try the effect of heating charcoal with the oxide of copper.

If this be done in a hard glass test-tube the copper oxide will be seen to change in appearance and copper to be obtained once more. If a test-tube containing lime water be held in a slanting position near the mouth of this test-tube as shown in Fig. 31, and the contents then shaken, the lime water will be found to turn milky, showing the presence of carbonic acid gas.

FIG. 31.

After this experiment the children will naturally wonder what would happen if charcoal were heated in the presence of the oxide of carbon, and this can be easily tried if some powdered charcoal is placed in a combustion tube, which is connected with an apparatus for generating carbonic acid gas on one side, and on the other with a flask containing a concentrated solution of caustic potash, as shown in Fig. 32.

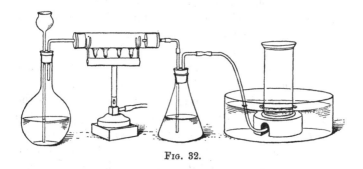

FIG. 32.

Carbon Monoxide. The air should all be driven out of the apparatus by passing a current of carbonic acid gas through it before the charcoal is heated, then when the charcoal is red-hot the collecting jar full of water can be inverted over the end of the delivery tube. When a jar full of the gas has been collected the children will want to put a lighted taper into it to see whether the light will be extinguished, and it is advisable for the teacher to warn the children to be careful in doing this. As the gas burns with a blue flame it is evidently not carbonic acid gas. Since the only substances present in the apparatus are carbonic acid gas and carbon, and carbonic acid gas has already been found to be a compound of carbon and oxygen, the new gas must be another compound of carbon and oxygen. If another jar of the gas be collected and some lime water poured into it and shaken up with the gas, there is no change : but if a light is applied and the cylinder covered with a glass plate and the contents shaken up the lime water will now be found to turn milky, showing the presence of carbonic acid gas. This new gas which when burnt forms carbonic acid gas is called "Carbon Monoxide," and carbonic acid gas is called carbon dioxide. The reasons for giving them these names cannot be understood at present and must be left for future investigation.

The children will easily realise that the blue flames seen on the top of a hot fire are due to the combustion of carbon monoxide. The oxygen of the air as it enters the grate combines with the carbon of the lower layers of the coal to form carbon dioxide: this passes over the red hot coal in the upper part of the grate and forms carbon monoxide which in its turn burns on the top of the fire and again produces carbon dioxide.

The children must be told that carbon monoxide is a very poisonous gas, and as large quantities of this gas are produced when coke and charcoal are burnt it constitutes a grave source of danger when these substances are burnt in badly ventilated rooms. The teacher will do well to point out the necessity of freeing the air of rooms from both carbon dioxide and carbon monoxide by thorough ventilation.

Carbon monoxide was first prepared by Lavoisier in 1777 by heating alum and charcoal together. Priestley showed in 1796 that this gas could be produced if iron scale (iron oxide) and powdered charcoal were heated together.

It has been found that, when copper was heated in air or in oxygen, copper oxide was formed, and when carbon was heated in a plentiful supply of air or in oxygen, carbon dioxide was formed; also that when lead was heated in air, first litharge, and then red lead, were

formed. The question now to be considered is, are
litharge and red lead oxides of lead ? To ascertain this,
pass a current of oxygen over finely divided lead heated
in a combustion tube: this lead will be seen to change
colour and the product when cool will resemble litharge.
This substance when heated with charcoal will give lead
again and carbon dioxide will be liberated.

It is known that red lead is produced when litharge is
heated for some time in air. Is the change from litharge
to red lead due to the union of the litharge with more

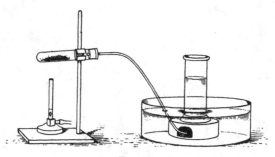

Fig. 33.

oxygen ? If it is, can this extra oxygen be driven off?
This question can easily be answered by heating some
red lead in a hard glass test-tube and collecting the gas
that is given off over water, using apparatus shown in
Fig. 33.

This gas will be found to be oxygen and litharge will
be left in the test-tube, so that red lead contains a larger
proportion of oxygen than litharge. Also if red lead be
heated with powdered charcoal, lead and carbon dioxide
are produced.

The children can try the effect of heating other metals

in the presence of air to see whether the oxides can be obtained. They will find that, when mercury is heated

Fig. 34.   Priestley.

in a test-tube, it always volatilises and condenses in small globules on the cool part of the tube : for this reason it is very difficult for them to obtain the oxide.   Red oxide of mercury can therefore be given to the children to investigate its properties. They will find that when it is heated in a test-tube, oxygen is given off and that mercury is left behind as a mirror on the sides of the test-tube, showing that the red oxide is simply a compound of mercury and oxygen.

Discovery of oxygen. It is of historic interest to know that oxygen was first obtained by Priestley in 1774 from oxide of mercury. He gives an account of his discovery in the following words:—

"Having produced a lens of 12 inches diameter and twenty inches focal distance, I proceeded with great alacrity to examine by the help of it what kind of air a great variety of substances, natural and fictitious, would yield.

On the 1st of August, 1774, I endeavoured to extract air from *mercurius calcinatus per se* and I presently found that, by means of this lens, air was expelled from it very readily. Having got about three or four times as much as the bulk of my material, I admitted water to it and found that it was not imbibed by it, but what surprised me more than I can well express was that a candle burned in this air with a remarkably vigorous flame."

Scheele, the Swedish chemist, gives the following account of his discovery of oxygen in his *Chemical treatise on Air and Fire* published in 1777:—

"I took a glass retort which was capable of holding 8 ounces of water and distilled fuming acid of nitre according to the usual method. In the beginning the acid went over red, then it became colourless and finally it all became red again; as soon as I perceived the latter I took away the receiver and tied on a bladder emptied of air into which I poured some thick milk of lime in order to prevent the corrosion of the bladder. I then proceeded with the distillation. The bladder began to expand gradually: after this I permitted everything to cool and tied up the bladder; lastly

I removed it from the neck of the retort. I filled a bottle which contained 10 ounces of water with this gas ; I then placed a small lighted candle in it. Scarcely had this been done than the candle began to burn with a large flame whereby it gave out such a bright light that it was sufficient to dazzle the eyes. I mixed one part of this air with three parts of that kind of air in which fire would not burn ; I had here an air which was like the ordinary air in every respect. Since this air is necessarily required for the origination of fire and makes up about the third part of our common air, I shall call it after this, for the sake of shortness, fire-air : but the other air which is not in the least serviceable for the fiery phenomenon, and makes up about two-thirds of our air, I shall designate after this with the name already known of Vitiated Air."

The following extract, giving an account of Lavoisier's discovery of oxygen, is from Robert Kerr's translation of Lavoisier's *Elements of Chemistry* :—

" I took a matrass of about 36 cubical inches capacity having a long neck of six or seven lines internal diameter, and having bent the neck to allow of its being placed in the furnace in such a manner that the extremity of its neck might be inserted under a bell glass placed in a trough of quicksilver. I introduced 4 ounces of pure mercury into the matrass, and by means of a syphon exhausted the air in the receiver so as to raise the quicksilver to another level, and I carefully marked the height at which it stood, by pasting on a slip of paper. Having accurately noted the height of the thermometer and barometer, I lighted a fire in the furnace, which I kept up almost continually during twelve days, so as to keep the quicksilver always very near its boiling point. Nothing remarkable took place during the first day. The mercury, though not boiling, was continually evaporating, and covered the interior surface of the vessel with small drops, which gradually augmenting to a sufficient size fell back into the mass at the bottom of the vessel. On the second day, small red particles began to appear on the surface of the mercury, these, during the four or five following days gradually increased in size and number, after which they ceased to increase in either respect. At the end of twelve days, seeing that the calcination of mercury did not at all increase, I extinguished the fire and allowed the vessel to cool. The bulk

FIG. 35. Lavoisier.

of air in the body and neck of the matrass and in the bell glass, reduced to a medium of 28 inches of the barometer and 54·5° of the thermometer, at the commencement of the experiment was about 50 cubical inches. At the end of the experiment the remaining air, reduced to the same medium pressure and temperature, was only between 42 and 43 cubical inches : consequently it had lost about ⅛ of its bulk. Afterwards having collected all the red particles, formed during the experiment, from the running mercury in which they floated, I found these to amount to 45 grains. I was obliged to repeat the experiment several times....

The air which remains after the calcination of the mercury in this experiment, and which was reduced to ⅚ of its former bulk, was no longer fit either for respiration or for combustion.

In the next place I took the 45 grains of red matter formed during this experiment, which I put into a small glass retort,

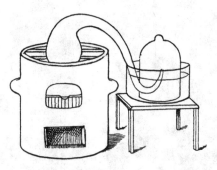

Fig. 36.

having a proper apparatus for receiving such liquid or gaseous product, as might be extracted. Having applied a fire to the retort in the furnace, I observed that, in proportion as the red matter became heated, the intensity of its colour augmented.

When the retort was almost red-hot the red matter began gradually to decrease in bulk and in a few minutes after it disappeared altogether : at the same time 41½ grains of running mercury were collected in the recipient, and 7 or 8 cubical inches of the elastic fluid, greatly more capable of supporting both respiration and combustion than atmospherical air, were collected in the bell glass.

This species of air was discovered almost at the same time by Dr Priestley, Mr Scheele and myself. Dr Priestley gave it the name of 'dephlogisticated air': Mr Scheele called it 'empyreal air': at first I named it 'highly respirable air,' for which has since been substituted the term of 'vital air.' In reflecting upon the circumstances of this experiment, we readily perceive that atmospheric air is composed of two elastic fluids of different and opposite qualities. As a proof of this important truth, if we recombine these two elastic fluids, which we have separately obtained in the above experiment, viz. the 42 cubical inches of mephites with the 8 cubical inches of highly respirable air, we reproduce an air precisely similar to that of the atmosphere, and possessing nearly the same power of supporting combustion and respiration and of contributing to the calcination of metals."

Manufacture of oxygen. Oxygen is now manufactured in large quantities, as it is used for many commercial purposes: for example, for the purification of coal from sulphur compounds, for the oxidation and thickening of oils for varnish: in conjunction with hydrogen or coal-gas it is used to produce the oxyhydrogen flame necessary for many purposes when a very high temperature is required.

Oxygen is also used to maintain the air in a respirable condition in places where it cannot be replaced, such as diving bells and submarine vessels; it is also used for medical purposes.

One of the cheapest methods of manufacturing oxygen on a large scale is that employed by the Brin Oxygen Company. Their process is based on the alternate formation and decomposition of barium peroxide. When baryta is heated to a dull red heat in a current of air, barium peroxide is formed; when this is heated more strongly it decomposes and oxygen is liberated. This process cannot go on indefinitely as baryta in a short time loses its power of absorbing oxygen, but it has been found that the permanency of the baryta is dependent upon the use of reduced pressure during deoxidation.

It has been found possible to dispense with change of temperature during the reaction, change of pressure being the means employed for the oxidation and deoxidation of the baryta.

This process was patented by M. Brin in 1880.

The children must now be told to refer to their question books to see whether this knowledge of oxygen

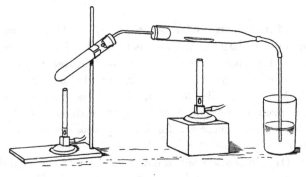

Fig. 37.

will enable them to solve any of their previous difficulties. At the beginning of the course they heated various substances in air, some of which charred and then practically disappeared.

The charred mass they will now realise consisted chiefly of carbon, which when heated combined with the oxygen of the air to form carbon dioxide and so seemed to disappear.

Sulphur, they found, burnt with a blue flame forming suffocating white fumes. Did the sulphur combine with the oxygen of the air, and were the white fumes an oxide of sulphur?

To answer these questions the children can pass a current of oxygen over heated sulphur, and allow the gas so formed to pass into water. The sulphur is found to burn with a dazzling blue flame in oxygen and the suffocating white fumes which are again formed are soluble in water.

The apparatus shown in Fig. 37 will be found convenient for this experiment. The solution obtained by dissolving the fumes in water will have an acid reaction on litmus, and if the petals of blue and violet flowers be dipped into the solution, the colour will be destroyed.

Sulphur dioxide. This suffocating gas is called sulphur dioxide, and the solution of the gas in water is called sulphurous acid.

Anhydride. The children can now be told that an oxide which enters into combination with water to form an acid is termed an "Anhydride," and therefore sulphur dioxide and carbon dioxide are both anhydrides, forming with water sulphurous acid and carbonic acid respectively. The children must now try to find out the other properties of sulphur dioxide. On account of its solubility in water they must collect this gas by downward displacement. They will find that the gas is non-inflammable and that a taper does not burn in it; also that it bleaches vegetable colouring matter, though not as readily as does chlorine.

If sulphur dioxide be passed into a solution of caustic soda it will be absorbed, and if the solution be left for some time crystals of sodium hydrogen sulphate will separate out. This substance is used by calico printers; as an antichlor for removing the last traces of chlorine from the bleached pulp used in paper-making; and also in photography.

Until a more convenient method of preparing sulphur dioxide has been discovered by the class, the further investigation of the sulphites should be postponed.

The oxides of carbon and sulphur are gaseous at ordinary temperature, and are soluble in water. The class can try whether the solid oxides known to them are also soluble in water. The simplest method of doing this is to put a little of the oxide in a test-tube, cover with distilled water, boil and filter and evaporate the filtrate to dryness on a watch-glass to see if there is any residue. It will be found that the oxides of copper, lead, and zinc are insoluble, while mercuric oxide and barium peroxide are slightly soluble.

CHAPTER IV.

THE children can now find out whether the metallic oxides they know are soluble in acids.

The quickest method of doing this is to place a row of small test-tubes in a stand and to put a small quantity of a different oxide into each test-tube, taking care to label the test-tube with the name of the oxide it contains. A watch-glass should be placed in front of each test-tube in which the solutions formed can be put to crystallise. Hydrochloric, sulphuric and nitric acids should in turn be tried.

(A) **Copper Chloride.** When *hydrochloric acid* is added to each of the oxides the following reactions should take place:—*Black copper oxide* will dissolve and form a green solution, which on dilution becomes blue, and from which rhombic prisms or needles of cupric chloride separate out. These crystals generally appear green on account of the adhering mother liquor, but in the pure state they are bright blue.

The probable causes of these changes of colour cannot at present be understood by the children. These crystals are very deliquescent, and on heating part with their water of crystallisation, becoming yellowish-brown in colour; on account of this property, a solution of copper chloride can be used as a sympathetic ink.

If the children try to write with a solution of copper chloride, using a clean steel pen for the purpose, they will find that the copper will be precipitated, and that therefore a glass rod or quill pen must be used.

Mercuric Chloride. *Red oxide of mercury* will dissolve in boiling hydrochloric acid, forming mercuric chloride, which crystallises from a hot solution in long, white, silky needles. Another name for mercuric chloride is corrosive sublimate. It is a powerful poison, for which the best antidote is white of egg. For this reason, unless the experiment can be directly supervised by the teacher, it should be omitted.

This substance has antiseptic properties, and is on this account largely used by taxidermists. It is also used for preserving timber from decay.

Lead Chloride. *Litharge* dissolves in hot hydrochloric acid, forming a solution from which on cooling, white, lustrous, needle-shaped crystals of lead chloride are formed.

Red lead will also dissolve in hydrochloric acid, forming the same compound, but in this case chlorine is evolved.

Zinc Chloride. Zinc oxide dissolves in hydrochloric acid, forming zinc chloride. If the solution be evaporated to the consistency of a syrup and a little concentrated hydrochloric acid be added, small deliquescent octahedrons separate out. This substance is used for mercerising cotton; it is also used in surgery as a caustic. It was first prepared by Glauber in 1648.

Barium Chloride. Barium peroxide dissolves in dilute hydrochloric acid, forming barium chloride. This substance will crystallise from the solution in colourless, flat, four-sided tables.

Fig. 38. Barium Chloride.

It is used for the preparation of "permanent white" (barium sulphate).

(B) While these solutions are crystallising, similar experiments can be tried in the same order, using *sulphuric acid* instead of hydrochloric acid.

Copper Sulphate. Copper oxide dissolves readily in sulphuric acid, forming a blue solution from which copper sulphate crystallises in large, blue triclinic prisms.

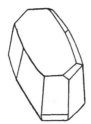

Fig. 39. Copper Sulphate.

This substance is also known as blue vitriol. When heated it loses its water of crystallisation and becomes white, but it regains its colour on the addition of water, and can therefore be used as a means of identifying water.

It is extensively used in calico printing: and in the preparation of Scheele's green, and pigments containing copper. It is also now used in some places for purifying the water in reservoirs.

Mercuric sulphate and lead sulphate, being insoluble in water, cannot easily be recognised by the children, or prepared by them by the action of sulphuric acid on the oxides.

Zinc Sulphate. *Zinc oxide* dissolves in sulphuric acid, forming a colourless solution from which, on concentration, crystals of zinc sulphate separate out. These crystals are colourless rhombic prisms.

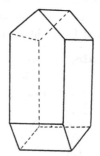

FIG. 40.   Zinc Sulphate.

This substance is sometimes called white vitriol. It is poisonous, and is used as a medicine; also in the manufacture of varnishes, and by dyers and calico printers.

Barium Sulphate. The children must be warned not to heat the mixture of *barium peroxide* and sulphuric acid, and the acid used must be very dilute. On the addition of this acid there is no change in appearance, but if the mixture be filtered, and the solid substance dried and then heated, no oxygen will be given off, showing that some chemical reaction has taken place.

If some of the white powder be heated with concentrated sulphuric acid and the liquid decanted off, bright silky needle-shaped crystals will separate out. This white powder is barium sulphate, and is largely used as a substitute for white lead in the manufacture of oil paints.

The filtrate can be shown by the teacher not to be sulphuric acid or pure water, for on being gently heated it evolves oxygen. The substance to which this is due is called hydrogen peroxide. It must be studied in greater detail at a later stage.

(C) The action of *nitric acid* on these oxides can now be tried in the same way.

Copper Nitrate. Copper oxide dissolves in nitric acid, and from this solution deep blue, prismatic, deliquescent crystals of copper nitrate separate out.

If these crystals be heated, red fumes are given off and the substance becomes black, and once more resembles copper oxide in appearance. If this black powder be heated with charcoal, copper is obtained and carbon dioxide given off, showing that this black powder is copper oxide. Thus they have arrived at a new method of preparing copper oxide. The red gas which was given off when the copper nitrate was heated will be more easily examined when another method of preparation has been discovered.

Mercuric Nitrate. *Red oxide of mercury* dissolves in nitric acid, forming a solution from which needle-shaped crystals of mercuric nitrate will separate out. When the crystals are heated a black powder is obtained, which on cooling becomes red and is recognised as mercuric oxide. So here again an oxide is obtained by the action of heat on a nitrate.

Lead Nitrate. *Litharge* dissolves in hot dilute nitric acid, forming a colourless solution from which milk-white regular octahedral crystals of lead nitrate separate out.

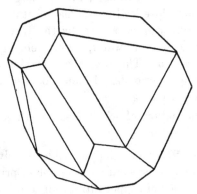

FIG. 41. Lead Nitrate.

When heated these crystals decrepitate and give off a red gas, a yellow deposit of litharge being left behind.

Lead nitrate is largely used in dyeing and calico printing, and for the preparation of chrome yellow.

Barium Nitrate. *Barium peroxide* dissolves in dilute nitric acid forming barium nitrate, which crystallises in large colourless anhydrous octahedrons.

When heated red fumes are given off, and barium oxide is left.

This substance is used in the preparation of green fire. If a little of it be heated on a platinum wire, it will colour the Bunsen flame green.

The children will realise from these experiments that some salts are formed by the action of acids on oxides, and some by the neutralisation of acids with alkalies. They will also realise that the oxides can be formed by

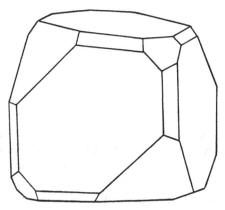

FIG. 42. Barium Nitrate.

heating the metals in air or in oxygen, and also by heating nitrates. At this point, they will naturally want to find out what will happen when an acid is poured on a metal.

CHAPTER V.

THE metals known to the children at this stage are copper, lead, zinc, tin, iron, mercury, silver and gold.

Mercury, silver and gold cannot very well be experimented with by a large class on account of the expense entailed, but if necessary the action of acids on these metals can be demonstrated by the teacher.

As in the case of the oxides, the different metals in a finely-divided condition should be placed in test-tubes which are labelled and put into a stand, and the acids poured upon each metal in turn.

If the action of hydrochloric acid be tried first on the metals, it will be found that no chemical reaction takes place when it is poured on copper, and that lead is only slowly attacked when heated with the hot concentrated acid. From the solution obtained, crystals of lead chloride separate out.

Zinc dissolves readily in this acid with brisk effervescence, showing that some gas is given off. The children will naturally test this gas by bringing a lighted taper to the mouth of the test-tube.

They will find that a slight explosion takes place, and that therefore a new gas is being evolved. When hydrochloric acid is poured upon tin and upon iron there is again effervescence, and the same inflammable gas is given off.

When all effervescence has ceased, the solution should be poured into watch-glasses and left to crystallise. Crystals of zinc chloride, tin chloride, and iron chloride separate out from the respective solutions.

Zinc chloride will be easily recognised from previous experiments.

Tin Chloride. Tin chloride, known as tin salt, crystallises in needles which are freely soluble in a small amount of water, but are apt to decompose when put into a large quantity, unless hydrochloric acid be present in excess. This substance is used as a mordant in dyeing and in calico printing.

Iron Chloride. From the solution of iron chloride it is possible to obtain crystals of a bluish-green colour, but they change on exposure to air and they are very deliquescent.

Hydrogen. While these solutions are crystallising, the properties of the new gas must be investigated. From their experiments the children will see that the new gas comes off most readily when zinc is acted upon by hydrochloric acid.

To obtain the gas in large quantities some granulated zinc can be placed in a Woulffe's bottle, fitted with two indiarubber corks, through one of which a thistle funnel passes, and through the other a delivery tube ending in a fine jet. The teacher must be careful that the apparatus used by each child is air tight. Some dilute hydrochloric acid is poured down the thistle funnel, and after effervescence has gone on for some minutes, a dry test-tube is inverted over the jet so that the end of the jet reaches to the top of the test-tube, and after a short interval is removed, held mouth downwards and a light applied. If the gas does not burn quite quietly, another test-tube full of the gas must be tested.

(It is well to warn the children that an interval must elapse before the same test-tube can be used for collecting more of the gas.)

They will find that if the test-tube be not held mouth downwards the gas will escape before it can be tested, showing that this gas is lighter than air.

When the gas in the test-tube burns quite quietly, a light can be applied to the end of the jet. The children have so far associated burning with the formation of an oxide, generally with carbon dioxide, but they will find that in this case a drop of lime water suspended from the convex surface of a watch-glass held over the flame does not turn milky, showing that carbon dioxide is not formed.

If however they hold a cold dry beaker over the flame, the inside of the beaker becomes misty, so that the substance formed is a liquid. When the flame is blown out the children will find that the liquid is still produced, though in a smaller quantity, and a little consideration will make them realise the reason of this and the necessity of placing a drying tube between the Woulffe's bottle and the jet. To collect more of this liquid, it is necessary to let the flame impinge upon a surface which can be kept cool.

The simplest way of doing this is to allow a stream of water to flow through a retort and to let the flame touch the curved surface of the retort, placing a small clean beaker underneath in which to collect the drops of the liquid which condense on the retort.

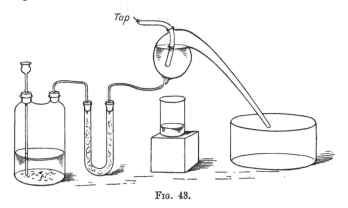

FIG. 43.

As soon as a little of the liquid has been collected, it can be tasted and tested with litmus. As it is neutral, tasteless, and colourless, the children will at once conclude that the liquid is water.

To confirm this view the freezing point must be determined, and this can easily be done by twisting a little cotton wool round the bulb of a thermometer, moistening it with the liquid, and putting the thermometer in a dry test-tube placed in a freezing mixture. When the liquid has solidified, take the thermometer out of the test-tube and note the temperature of fusion.

The boiling point can be found by putting some of the liquid in a test-tube, boiling the liquid and holding the bulb of the thermometer in the steam until the mercury remains stationary for some time, which it will be found to do at 100° C. Another simple and confirmatory test is to add this liquid to anhydrous copper sulphate, which will turn blue, and the children have previously found that this change only takes place on the addition of water. The liquid formed when this new gas burns in air is therefore water, and on this account the name "Hydrogen" has been given to the gas.

Is water an oxide of hydrogen? To obtain an answer to this question, dry hydrogen gas can be passed over dry heated copper oxide.

A thin layer of dry copper oxide can be placed in a porcelain boat in a combustion tube. The hydrogen can be generated in a Woulffe's bottle between which and the combustion tube a drying tube containing calcium chloride is placed. At the other end of the combustion tube is a delivery tube leading into a test-tube surrounded by cold water.

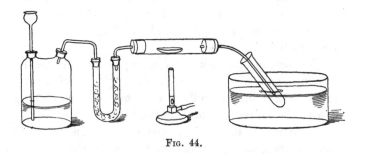

Fig. 44.

For class work it is advisable to have the apparatus as small and as simple as possible to eliminate the possibility of accidents, but the teacher must be careful to go round the class and see that every apparatus is air-tight, and warn the children not to remove any cork after the experiment has been begun.

Before heating the copper oxide all the air in the apparatus must be displaced by hydrogen. When the copper oxide is heated drops of moisture will condense in the test-tube and the substance in the boat will become red in colour. On examination of the substances formed, the liquid will be found to be water and the red substance in the boat to be copper.

Composition of Water by Weight. This experiment must next be tried quantitatively. The combustion tube and empty porcelain boat must first be weighed, then copper oxide put in the boat and the whole reweighed: the difference between the two weighings will give the weight of the copper oxide used. The delivery tube must be replaced by a weighed calcium chloride tube and the experiment conducted as before. At the end of the experiment the combustion tube and contents must be again weighed and the loss in weight will give the amount of oxygen that has combined with the hydrogen. The gain in weight of the calcium chloride tube gives the amount of water formed, and from these results the proportion by weight in which hydrogen and oxygen combine to form water can be calculated.

The careful workers in the class can substitute a Kipp's hydrogen apparatus for the Woulffe's bottle, and the gas can be dried by bubbling it slowly through concentrated sulphuric acid, as this method gives more accurate results, but is hardly suitable for general class work.

NOTE. (Paracelsus in the 16th century obtained an inflammable gas by the action of dilute acids on certain metals. The true nature of this gas was first demonstrated by Cavendish in 1766 and was called by him "inflammable air." Cavendish also discovered in 1783 that water was a compound of hydrogen and oxygen and his discovery was confirmed in the same year by Lavoisier and Laplace.)

The effect of passing hydrogen over the other known oxides, when heated, can be tried by substituting in turn porcelain boats containing various oxides for the one containing copper oxides: and in addition to the known oxides the puce-coloured powder which was obtained when red lead was treated with dilute nitric acid can be placed in one boat, and the children will discover in the course of this experiment that this substance is another oxide of lead.

It will be found that hydrogen will reduce zinc oxide, lead oxide, and iron oxide to the metallic state, provided that the hydrogen is passed over the oxides rapidly, but that the metals, manganese and barium, are not obtained by passing hydrogen over the heated oxides of these metals, although water is produced in each case. If time permits, these experiments should be done quantitatively, and thus the combining weights of the metals with oxygen will be found, and it will be discovered that lead combines with oxygen in different proportions.

Although hydrogen is a constituent of water, none of this gas is evolved when water is poured upon cold zinc, iron, or tin: yet hydrogen is evolved when dilute hydrochloric acid is poured on these metals. The children will therefore think it probable that hydrogen is a constituent of hydrochloric acid, and they will try to find out whether this is true by passing dry hydrochloric acid gas over one of these metals. If iron be selected it will be found that when cold there is no reaction, but when the iron is heated a change takes place and hydrogen is given off.

The gas can be collected over water: any unchanged hydrochloric acid gas which passes over will dissolve in the water. The substance left in the tube is iron chloride, and will easily be recognised by its properties.

Constitution of Hydrochloric Acid Gas. Hydrogen therefore is a constituent of hydrochloric acid gas. When hydrochloric acid was poured upon red lead chlorine was evolved, and as red lead is an oxide, the chlorine must have come from the acid. Hydrochloric acid gas seems therefore to be a compound of hydrogen and chlorine. To confirm this supposition a jet of

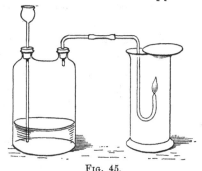

FIG. 45.

hydrogen should be burned in a jar of chlorine. This experiment should be tried only by selected members of the class or by the teacher.

It can most easily be done by lowering a lighted jet of hydrogen into a jar full of chlorine, covering the jar as much as possible with a glass plate.

After the flame has gone out, the product in the jar can be tested with litmus, and by bringing a rod dipped in ammonia to the mouth of the jar; thus it can be shown to be hydrochloric acid gas.

As hydrochloric acid gas is a compound of hydrogen and chlorine, will water be formed when the gas is passed over a heated metallic oxide?

This experiment can be tried if the apparatus shown in Fig. 44 be used, only rock salt and sulphuric acid must be placed in the Woulffe's bottle instead of the zinc and hydrochloric acid.

If iron oxide be placed in the boat in the combustion tube, it will be found that iron chloride will be left in the boat, and in the test-tube there will be water acidulated with hydrochloric acid. When zinc oxide is placed in the boat, zinc chloride will be formed and a solution of hydrochloric acid.

Thus these chlorides are found to be compounds of these metals and chlorine. Are the other chlorides known to the class also compounds of a metal and chlorine? Is the calcium chloride which has so often been used as a drying agent a compound of a metal calcium and chlorine?

Calcium chloride has already been prepared by treating lime with hydrochloric acid, and chlorides have been prepared by treating oxides with hydrochloric acid. Is lime therefore an oxide of calcium?

If lime is placed in the boat in the combustion tube and a current of dry hydrochloric acid gas passed over the heated lime, water acidulated with hydrochloric acid is again obtained and calcium chloride is left in the boat, thus showing that lime behaves like an oxide. The class can be told that it is the oxide of calcium, a metal which is difficult to isolate. A specimen of calcium can be shown to the children.

The class will now be able to solve one of the

difficulties which they came across in the earlier part of their work, and that is the composition of chalk.

When chalk was heated carbon dioxide was given off and lime was left: since lime is an oxide of calcium, chalk is a compound formed by the union of two oxides, carbon dioxide an acid oxide, and lime a basic oxide.

Calcium carbonate = Calcium oxide + carbon dioxide.

Thus chalk contains the elements calcium, carbon and oxygen.

When barium peroxide was heated, oxygen was evolved and a white powder which resembles lime was left. This substance can be shown to be slightly soluble in water, and like lime it forms an alkaline solution. If hydrochloric acid gas be passed over this substance when heated, a solution of hydrochloric acid is formed, showing that water has been produced, and a white substance is left, which on examination is found to be barium chloride: so that the white powder produced when barium peroxide is heated is another oxide of barium : this substance is usually called baryta.

In Black's researches on Magnesia Alba a substance was obtained which resembled lime : this substance can also be shown to be an oxide of a metal which cannot be isolated by the class and which is called Magnesium. The children can be given strips of magnesium ribbon to burn and they will find that they obtain a white oxide. This experiment can be done quantitatively also.

The use of magnesium in flash light photography can be told to the class.

It has been shown that iron chloride, zinc chloride, barium chloride and calcium chloride are all compounds of a metal and chlorine. Are the substances which were named sodium chloride and potassium chloride also compounds of a metal and chlorine? When calcium carbonate was acted upon by hydrochloric acid, calcium chloride and carbon dioxide were produced. When sodium carbonate was acted upon by hydrochloric acid a similar reaction took place, carbon dioxide was given off and sodium chloride was formed. If some washing-soda be heated till all the water of crystallisation is given off, and then put into a porcelain boat in the combustion tube used in the previous experiments, and a stream of dry hydrogen chloride passed over the heated substance, carbon dioxide will be given off, a solution of hydrochloric acid will be formed, and sodium chloride, which can easily be identified, will be left in the boat.

Hydrogen chloride + Sodium carbonate = Sodium chloride + carbon dioxide + water.

So that it seems probable that sodium chloride is a compound of a metal and chlorine, and that sodium carbonate is a compound of the oxide of this metal and carbon dioxide.

The class can be told that sodium is another metal which they cannot themselves isolate. A specimen of sodium can be shown to them and they can themselves prove that sodium chloride is a compound of the metal and chlorine by collecting some dry hydrochloric acid gas in a test-tube over mercury and introducing into it a pellet of sodium on the end of a bent wire. The mercury will be seen to rise in the test-tube until the gas only occupies half of its original volume; and the sodium will be covered with a white incrustation, which can easily be shown to be sodium chloride, while the gas left in the test-tube is hydrogen. This experiment also shows that hydrogen chloride contains half its own volume of hydrogen.

Decomposition of Water by Metals. When sodium is placed in water a brisk reaction takes place. A pellet of sodium about the size of a split pea should be placed on a piece of wire gauze and the gauze folded over the sodium with a pair of crucible tongs: when this is dropped into water in a trough and a test-tube full of water inverted over it, bubbles of gas are seen to rise and displace the water in the test-tube. When heated, this gas is found to be hydrogen and the liquid in the trough to have an alkaline reaction to litmus, while the fingers feel soapy when dipped into it. As hydrogen has been given off during the reaction it is evident that the metal sodium has decomposed the water.

Will any other metals decompose water? The children know that water at ordinary temperatures is not decomposed by the common metals, but they can try whether steam is acted upon by these metals when heated. This can be done by placing the metal in a porcelain lined iron tube and heating it in a current of steam.

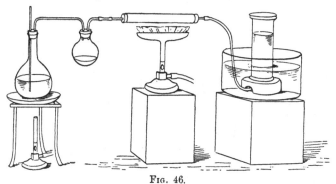

Fig. 46.

Bright iron nails can first be placed in the tube and strongly heated ; when they are red-hot the current of steam can be passed over them and the end of the delivery tube placed under a jar full of water inverted over a beehive shelf standing in a trough of cold water. Bubbles of gas will be seen to rise and displace the water. This gas on testing will be found to be hydrogen, while the iron nails are coated with iron oxide, showing that steam is decomposed by red-hot iron.

If copper filings be substituted for the iron nails no change will take place. When granulated zinc is put in the tube and heated, steam is again decomposed. These experiments show that some metals have a greater affinity for oxygen than others.

Hydrogen has been found to be a constituent of water and of hydrochloric acid. Is it also a constituent of sulphuric acid and of nitric acid ? The children can try the effect of pouring dilute and concentrated sulphuric acid on the metals, iron, zinc, tin, magnesium and copper. They will find that the metals iron, zinc, tin and magnesium act upon dilute sulphuric acid with the evolu-

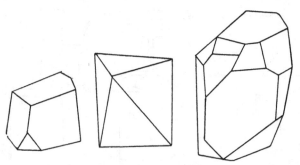

FIG. 47.   Iron Sulphate Crystals.

tion of hydrogen and the formation of the sulphates of these metals.

Iron Sulphate. The solution obtained by dissolving iron filings in dilute sulphuric acid has a green colour, and when left to evaporate, green monoclinic crystals of ferrous sulphate separate out.

This substance will be recognised to be the same as that from which sulphuric acid was originally obtained, and which the pupils now realise to be a compound of iron.

Ferrous sulphate is used largely in the manufacture of Prussian blue and of ink.

Zinc Sulphate. Zinc sulphate they will now readily recognise from its characteristic crystals.

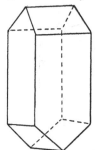

Fig. 48. Zinc Sulphate.

Tin dissolves in dilute sulphuric acid, forming a colourless solution from which on concentration crystals of basic stannous sulphate separate out.

Magnesium Sulphate. Magnesium dissolves in dilute sulphuric acid and crystals of magnesium sulphate separate out, from a hot concentrated solution, in four-sided rhombic prisms isomorphous with zinc sulphate. These crystals, which are commonly known as Epsom salts, are used medicinally and also by dyers.

Dilute sulphuric acid has no action on copper.

Since hydrogen is evolved when sulphuric acid is acted upon by some metals, sulphuric acid must contain hydrogen. Is the same amount of hydrogen evolved from sulphuric acid as from hydrochloric acid when acted upon by the same weight of the metal ?

To answer this question ·2 gram of magnesium should be weighed out. A small flask is fitted with a cork, through which passes a tube bent at right angles and connected with a Winchester full of water. The mouth of the Winchester is closed by a two-holed cork, fitted with tubes as in Fig. 49.

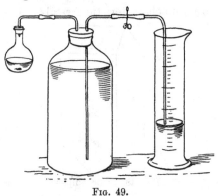

FIG. 49.

Some dilute hydrochloric acid is placed in the flask, a piece of magnesium ribbon is dipped into it and the cork quickly replaced. The hydrogen which is evolved displaces the water in the Winchester and this water is collected in a graduated cylinder. The volume of this water will give the volume of hydrogen evolved from hydrochloric acid when acted upon by ·2 gram of magnesium.

The same apparatus can be used in the next experiment when dilute sulphuric acid is substituted for hydrochloric acid, and the same weight of magnesium is used. It will be found that the same volume of hydrogen is evolved.

The same experiments can be repeated, using zinc instead of magnesium, and the class will discover that again hydrochloric acid and sulphuric acid yield the same amount of hydrogen when acted upon by the same weight of zinc, but that the volume of hydrogen liberated depends on the metal used.

When trying these experiments it must be remembered that during the reaction the small flask becomes hot, and that therefore too much water is forced out of the Winchester, and this would give too high a result. To avoid this error the tube leading into the graduated cylinder must not be taken out of the water in the cylinder until the flask is cool, and then care must be taken that the water is at the same level in the Winchester as in the cylinder when the reading is taken.

To find the weight of hydrogen evolved when sulphuric acid is acted upon by zinc, a thin glass flask is fitted with a two-holed cork through which pass two pieces of glass tubing bent at right angles : the end of one piece is closed by means of indiarubber tubing and a clamp. The other glass tube is connected with a U-tube containing calcium chloride. Some dilute sulphuric acid is placed in the flask and the whole apparatus weighed. One gram of zinc foil is dropped into the flask and the cork quickly replaced. After the effervescence has entirely ceased, air must be drawn through the apparatus, which can then be reweighed and the loss in weight will give the weight of hydrogen obtained from sulphuric acid by the action of 1 gram of zinc.

FIG. 50.

Weight of a Litre of Hydrogen.   Since the volume of hydrogen given off from sulphuric acid by a known weight of zinc has been found, the volume which would be given off by 1 gram of zinc can be calculated. Since the weight of hydrogen given off when a gram of zinc acts upon sulphuric acid is known, the weight of a litre of hydrogen can easily be calculated.

Sulphur Dioxide.   Concentrated sulphuric acid has no action upon iron, zinc, or magnesium; but when heated with copper, effervescence takes place and a suffocating gas is given off, leaving a dark-coloured residue in the flask.   If this be filtered a blue liquid is obtained, from which, on evaporation, crystals of copper sulphate separate.   The dark-coloured residue can be kept for future investigation.

If the gas which is evolved be passed into water, an acid solution is obtained, which bleaches vegetable colouring matter.

The gas will be recognised as sulphur dioxide. As it has been obtained by the action of the element copper on sulphuric acid, and as sulphur dioxide is known to consist of the elements sulphur and oxygen, sulphuric acid must contain sulphur and oxygen, and it is also known to contain hydrogen.

Sulphurous acid is a compound of sulphur dioxide and water. Sulphuric acid contains the elements sulphur, oxygen and hydrogen. If it be a compound of these elements only, it seems probable that it is a compound of water and another oxide of sulphur. If there is another oxide of sulphur it must contain more or less oxygen than sulphur dioxide.

Carbon monoxide was formed by passing carbon dioxide over heated carbon. Does any change take place when sulphur dioxide is passed over heated sulphur ? This experiment can be tried by the class, and they will find that sulphur will not burn in a current of sulphur dioxide and that the sulphur dioxide has not changed. Can sulphur dioxide be made to combine with more oxygen ?

The class must here be told that it has been discovered by experiments that substances will sometimes combine in the presence of another substance, which seems unaltered at the end of the experiment. Sulphur dioxide and oxygen can be made to combine when passed over heated platinised asbestos. Asbestos is a familiar substance, as the little mats in the laboratory are made of it. Platinum the pupils have often used in the form of wire and foil.

Platinised asbestos consists of asbestos coated with finely divided platinum. This substance can be prepared by dipping some ignited asbestos into platinic chloride, and then into ammonium chloride solutions, and igniting in the Bunsen flame. This should be done by the teacher.

Sulphur Trioxide. The oxygen and sulphur dioxide must be dried before they are passed over the platinised asbestos in the combustion tube. This is most easily done by passing a current of oxygen and sulphur dioxide through a flask containing concentrated sulphuric acid. The mixture of the dry gases is then passed over the heated platinised asbestos and the substance formed collected in a dry **U**-tube standing in a freezing mixture.

The apparatus can be set up as in Fig. 51, or a

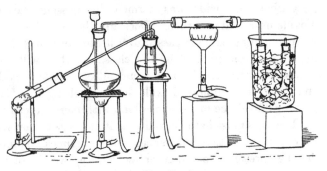

FIG. 51.

Winchester containing oxygen and connected with the water-tap can be substituted for the test-tube containing potassium chlorate.

Care must be taken that the combustion tube and the **U**-tube are perfectly dry. The oxygen can be prepared by heating a mixture of potassium chlorate and manganese dioxide in a test-tube. The sulphur dioxide by heating copper turnings and concentrated sulphuric acid. White silky crystals will form in the U-tube. These crystals will be found to combine with water with great energy and form sulphuric acid. These white crystals are therefore the anhydride of sulphuric acid, and are known by the name of sulphur trioxide.

The class may think that sulphur trioxide does not consist of sulphur and oxygen only, as the sulphur dioxide and oxygen have been passed over platinised asbestos. They can easily prove that this substance has not been changed by the reaction, if they weigh the tube containing platinised asbestos before and after the experiment. They will find that the weight remains the same.

So they have proved that sulphuric acid contains only the elements sulphur, hydrogen, and oxygen.

The following equations will be useful as a summary of this work:

Sulphur + Oxygen   = Sulphur dioxide.
Sulphur dioxide + Water = Sulphurous acid.
Sulphur dioxide + Oxygen = Sulphur trioxide.
Sulphur trioxide + Water = Sulphuric acid.

From the action of metals on hydrochloric acid and on sulphuric acid the elements present in these acids have been discovered. Will the action of metals on nitric acid lead to a similar result? When dilute nitric acid is poured upon magnesium, hydrogen is evolved, as it was when the other acids were poured upon it, showing that hydrogen is a constituent of nitric acid.

When nitric acid is poured upon copper a red gas is given off. The properties of this new gas must be investigated. Some copper turnings can be placed in a flask fitted with a thistle funnel and delivery tube passing through a two-holed cork, the end of the delivery tube dipping under water contained in a pneumatic trough. Some concentrated nitric acid is mixed with an equal volume of water and poured through the thistle funnel.

A brisk reaction will be found to take place and the flask will at first be filled with red fumes ; but in a short time the gas in the flask is colourless. A cylinder full of water can now be inverted over the end of the delivery tube and the water will be displaced by a colourless gas.

If a jar full of the gas be exposed to the air the contents at once become reddish-brown in colour.

If a lighted taper be plunged into a jar of the gas it at once goes out and the gas does not burn.

The children can try whether other substances such as sulphur and magnesium which readily burn in air will continue to burn in this gas. They will find that the magnesium continues to burn, but the sulphur goes out. The next problem will be to find what substances are formed when magnesium burns in this gas.

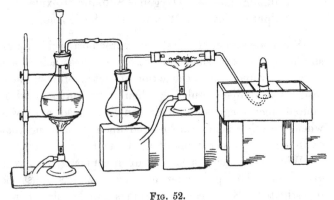

Fig. 52.

The dried gas can be passed over magnesium wire in a combustion tube, to which is connected a delivery tube leading into a trough containing water as in Fig. 52.

When all the air has been displaced and the gas in the apparatus is colourless, the magnesium can be heated, and in a short time will be found to burn brilliantly. If the gas which is now issuing from the delivery tube be collected in a test-tube and tested, it will be found to be nitrogen. So that this experiment gives a new method of preparing nitrogen.

The white substance left in the combustion tube will be recognised as magnesium oxide, showing that the gas contains oxygen and nitrogen.

Nitric Oxide. This gas is called nitric oxide. Priestley was the first to fully investigate this gas and he called it "Nitrous Air."

A cylinder full of nitrogen can be collected by this method and the nitrogen used to find out whether the change in colour of nitric oxide when exposed to the air is due to the oxygen or to the nitrogen of the air. Nitrogen will be found to produce no change, while if oxygen be slowly bubbled into a jar of the gas over water the colour of the gas will change and the water will rise in the jar. This experiment shows that the red gas is another oxide of nitrogen which is soluble in water, yielding an acid solution.

Nitrogen Peroxide. The name given to it is "Nitrogen Peroxide." Nitric oxide can therefore be used to detect the presence of oxygen.

Since nitric oxide is a compound of nitrogen and oxygen, and is obtained from nitric acid by the action of copper on this acid, nitric acid must contain nitrogen and oxygen. It has already been shown to contain hydrogen, therefore the elements hydrogen, oxygen and nitrogen are present in nitric acid.

If the mixture left in the flask used in the preparation of nitric oxide be filtered, and the filtrate evaporated and left to crystallise, deep blue crystals separate out which will be recognised as the same as those obtained when copper oxide was treated with nitric acid.

The red gas which is given off when the crystals of copper nitrate are heated can now be identified as nitrogen peroxide.

Nitrates. The nitrates known to the children are the following :—potassium nitrate, sodium nitrate, ammonium nitrate, calcium nitrate, lead nitrate, barium nitrate and mercuric nitrate.

They have already found out that when potassium nitrate is heated it fuses, and if a piece of charcoal is dropped into it, the charcoal takes fire and burns brightly.

They will now think it probable that this reaction is due to the evolution of oxygen.

Some potassium nitrate should be put in a test-tube, fitted with a cork and delivery tube. When the contents of the test-tube are heated, a gas is evolved which can be collected over water and shown to be oxygen.

The residue in the test-tube cannot now be potassium nitrate. If dilute sulphuric acid be added to it, red fumes will be given off and potassium sulphate will be left. The red fumes will be recognised as nitrogen peroxide, showing that the residue in the test-tube contained nitrogen and oxygen. As oxygen is evolved when potassium nitrate is heated the residue must contain less oxygen. This substance is known as potassium nitrite.

When sodium nitrate is heated a similar reaction takes place. Ammonium nitrate is a substance which the children will remember, as its behaviour when heated was so peculiar. They will know that great care must be taken when heating it. As it is hardly wise to let a class of children experiment to find out the best means of collecting the gas, the teacher can give directions for its preparation.

A small flask is fitted with a one-holed cork through which passes a glass tube bent twice at right angles, and leading into a Woulffe's bottle, from which a delivery tube passes into a trough containing warm water.

Fig. 53.

Some ammonium nitrate is placed in the flask (the layer of ammonium nitrate must be at least one inch in depth) and heated very gently until the salt fuses and forms a colourless liquid. When bubbles of gas begin to rise in the trough full of warm water a cylinder full of warm water must be inverted over the delivery tube and the gas collected.

When two or three jars of this gas have been collected, a flask should be filled with warm water and fitted with a cork and two tubes bent at right angles as in Fig. 54.

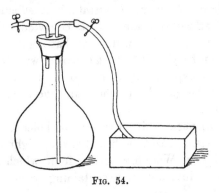

Fig. 54.

The gas must be passed into this flask until all the water has been displaced, and the clamps then closed.

The gas in the collecting jars can then be tested. It will be found to support the combustion of a glowing splint, but if nitric oxide be decanted into a jar of this gas, red fumes will not be formed, showing that this gas is not oxygen.

Sulphur when placed in a deflagrating spoon and strongly heated will continue to burn in the gas with the production of sulphur dioxide. So this gas must contain oxygen.

The flask full of the gas must now be connected to a small combustion tube containing magnesium wire, and the apparatus fitted up as in Fig. 55.

A wash bottle containing concentrated sulphuric acid must be placed between the flask and the combustion tube to dry the gas.

The apparatus must be filled with the gas and then the magnesium heated. The magnesium takes fire and burns brilliantly, forming oxide of magnesium, and nitro-

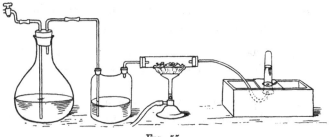

Fig. 55.

gen collects in the test-tube, so this gas, like nitric oxide and nitrogen peroxide, is an oxide of nitrogen.

Nitrous Oxide. The name given to this gas is "Nitrous Oxide."

It will be found that this gas is soluble in cold water. Nitrous oxide is an anaesthetic known by the name of "Laughing gas." These three oxides of nitrogen can easily be distinguished, as nitric oxide is a colourless gas which turns red on exposure to air, nitrogen peroxide is a red gas soluble in water, and nitrous oxide is a colourless gas which supports combustion but does not combine with nitric oxide to form a red gas.

At the end of this course the children will know the characteristic properties of a salt, and that a salt can be obtained by the neutralisation of an acid by an alkali, by the action of an acid on an oxide, by displacement of the hydrogen of an acid by a metal, or by double decomposition.

They will also to some extent understand what is meant by an acid, an alkali, an acid oxide, a basic oxide, and an anhydride.

In addition to this, when the results of their experiments are collected and summarised the following conclusions will be arrived at:—

1.   Carbon combines with oxygen in two proportions, forming the compounds called carbon monoxide and carbon dioxide.

2.   Sulphur combines with oxygen in two proportions, forming the compounds called sulphur dioxide and sulphur trioxide.

3.   Nitrogen combines with oxygen in at least three proportions as, although the anhydride of nitric acid has not been prepared by the class, they have obtained the compounds called nitrous oxide, nitric oxide, and nitrogen peroxide.

4.   Hydrogen combines with oxygen in two proportions, forming the compounds known as water and hydrogen peroxide.

5.   Lead combines with oxygen in three proportions, forming the compounds known as litharge, red lead and lead peroxide.

6.   Barium combines with oxygen in two proportions, forming the compounds known as baryta and barium peroxide.

7.   Carbon dioxide combines with sodium hydroxide, with potassium hydroxide and with calcium hydroxide in two proportions.

It has also been found that the same weight of oxygen enters into combination with different weights of hydrogen, carbon, copper and magnesium to form water, carbon dioxide, copper oxide, and magnesium oxide; but that in the same compound the elements present in it are always combined in the same proportion by weight.

It will thus be seen that these few experiments lay the foundation of a correct understanding of Dalton's " laws " of definite and multiple proportions, and form an introduction to Theoretical Chemistry.

INDEX.

Printed in the United States
By Bookmasters